Mathematical
Surveys
and
Monographs

Volume 113

Homotopy Limit Functors on Model Categories and Homotopical Categories

William G. Dwyer
Philip S. Hirschhorn
Daniel M. Kan
Jeffrey H. Smith

American Mathematical Society

EDITORIAL COMMITTEE

Jerry L. Bona   Peter S. Landweber, Chair
Michael G. Eastwood   Michael P. Loss
J. T. Stafford

2000 *Mathematics Subject Classification.* Primary 18A99, 18D99, 18G55, 55U35.

For additional information and updates on this book, visit
www.ams.org/bookpages/surv-113

Library of Congress Cataloging-in-Publication Data

Homotopy limit functors on model categories and homotopical categories / William G. Dwyer ... [et al.].
     p. cm. — (Mathematical surveys and monographs, ISSN 0076-5376; v. 113)
  Includes bibliographical references and index.
  ISBN 0-8218-3703-6 (alk. paper)
  1. Homotopy theory.   I. Dwyer, William G., 1947–   II. Series.

QA612.7.H635 2004
514′.24—dc22   2004059481

   Copying and reprinting. Individual readers of this publication, and nonprofit libraries acting for them, are permitted to make fair use of the material, such as to copy a chapter for use in teaching or research. Permission is granted to quote brief passages from this publication in reviews, provided the customary acknowledgment of the source is given.
   Republication, systematic copying, or multiple reproduction of any material in this publication is permitted only under license from the American Mathematical Society. Requests for such permission should be addressed to the Acquisitions Department, American Mathematical Society, 201 Charles Street, Providence, Rhode Island 02904-2294, USA. Requests can also be made by e-mail to reprint-permission@ams.org.

   © 2004 by the American Mathematical Society. All rights reserved.
      The American Mathematical Society retains all rights
      except those granted to the United States Government.
         Printed in the United States of America.
   ⊚ The paper used in this book is acid-free and falls within the guidelines
         established to ensure permanence and durability.
         Visit the AMS home page at http://www.ams.org/
            10 9 8 7 6 5 4 3 2 1   09 08 07 06 05 04

To

Sammy Eilenberg

Dan Quillen

Pete Bousfield

Contents

Preface   vii

Part I. Model Categories   1

Chapter I. An Overview   3
  1. Introduction   3
  2. Slightly unconventional terminology   3
  3. Problems involving the homotopy category   5
  4. Problem involving the homotopy colimit functors   8
  5. The emergence of the current monograph   11
  6. A preview of part II   12

Chapter II. Model Categories and Their Homotopy Categories   19
  7. Introduction   19
  8. Categorical and homotopical preliminaries   22
  9. Model categories   25
  10. The homotopy category   29
  11. Homotopical comments   32

Chapter III. Quillen Functors   35
  12. Introduction   35
  13. Homotopical uniqueness   38
  14. Quillen functors   40
  15. Approximations   42
  16. Derived adjunctions   44
  17. Quillen equivalences   48
  18. Homotopical comments   51

Chapter IV. Homotopical Cocompleteness and Completeness of Model Categories   55
  19. Introduction   55
  20. Homotopy colimit and limit functors   59
  21. Homotopical cocompleteness and completeness   62
  22. Reedy model categories   65
  23. Virtually cofibrant and fibrant diagrams   69
  24. Homotopical comments   72

Part II. Homotopical Categories   75

Chapter V. Summary of Part II   77
  25. Introduction   77

26.	Homotopical categories	78
27.	The hom-sets of the homotopy categories	80
28.	Homotopical uniqueness	82
29.	Deformable functors	83
30.	Homotopy colimit and limit functors and homotopical ones	85

Chapter VI. Homotopical Categories and Homotopical Functors ... 89
31.	Introduction	89
32.	Universes and categories	93
33.	Homotopical categories	96
34.	A colimit description of the hom-sets of the homotopy category	101
35.	A Grothendieck construction	103
36.	3-arrow calculi	107
37.	Homotopical uniqueness	112
38.	Homotopically initial and terminal objects	115

Chapter VII. Deformable Functors and Their Approximations ... 119
39.	Introduction	119
40.	Deformable functors	123
41.	Approximations	126
42.	Compositions	130
43.	Induced partial adjunctions	133
44.	Derived adjunctions	138
45.	The Quillen condition	143

Chapter VIII. Homotopy Colimit and Limit Functors
and Homotopical Ones ... 147
46.	Introduction	147
47.	Homotopy colimit and limit functors	148
48.	Left and right systems	152
49.	Homotopical cocompleteness and completeness (special case)	159
50.	Homotopical colimit and limit functors	161
51.	Homotopical cocompleteness and completeness (general case)	166

Index ... 171

Bibliography ... 181

Preface

This monograph, which is aimed at the graduate level and beyond, consists of two parts.

In part II we develop the beginnings of a kind of "relative" category theory of what we will call *homotopical categories*. These are categories with a single distinguished class of maps (called *weak equivalences*) containing all the isomorphisms and satisfying one simple *two out of six* axiom which states that

(∗) for every three maps r, s and t for which the *two* compositions sr and ts are defined and are weak equivalences, the *four* maps r, s, t and tsr are also weak equivalences,

which enables one to define "homotopical" versions of such basic categorical notions as initial and terminal objects, colimit and limit functors, adjunctions, Kan extensions and universal properties.

In part I we use the results of part II to get a better understanding of Quillen's so useful *model categories*, which are categories with three distinguished classes of maps (called *cofibrations*, *fibrations* and *weak equivalences*) satisfying a few simple axioms which enable one to "do homotopy theory". In particular we show that such model categories are *homotopically cocomplete* and *homotopically complete* in a sense which is much stronger than the existence of all small homotopy colimit and limit functors.

Both parts are essentially self-contained. A reader of part II is assumed to have some familiarity with the categorical notions mentioned above, while those who read part I (and especially the introductory chapter) should also know something about model categories. In the hope of increasing the local as well as the global readability of this monograph, we not only start each section with some introductory remarks and each chapter with an introductory section, but also each of the two parts with an introductory chapter, with the first chapter of part I serving as motivation for and introduction to the whole monograph and the first chapter of part II summarizing the main results of its other three chapters.

Part I

Model Categories

CHAPTER I

An Overview

1. Introduction

1.1. Summary. This monograph is essentially the result of an unsuccessful attempt to give an updated account of Quillen's *closed model categories*, i.e. categories with three distinguished classes of maps (called *weak equivalences, cofibrations and fibrations*) satisfying a few simple axioms which enable one to "do homotopy theory". That attempt however failed because (see 25.1), the deeper we got into the subject, and especially when we tried to deal with homotopy colimit and limit functors on arbitrary model categories, the more we realized that we did not really understand the role of the *weak equivalences*.

In this introductory chapter we will therefore try to explain the problems we ran into and the solutions we came up with, and how all this determined the content and the layout of this monograph.

1.2. Organization of the chapter. After fixing some slightly unconventional terminology (in §2), we discuss the problems we encountered in dealing with the *homotopy category* of a model category (in §3) and with the *homotopy colimit functors* on a model category (in §4) and explain (in §5) how all this led to the current two-part monograph. In the first part we review some mostly known results on *model categories*, but do so whenever possible from the "homotopical" point of view which we develop in the second part, and in the second part we investigate what we call *homotopical categories* which are categories with only a single distinguished class of maps (called *weak equivalences*). The last section (§6) then contains some more details on this second part.

2. Slightly unconventional terminology

In order to be able to give a reasonably clear formulation of the above mentioned problems and their solutions, we slightly modify the customary meaning of the terms "model category" and "category" and introduce, just for use in this introductory chapter, the notion of a "category with weak equivalences" or "we-category".

2.1. Model categories. In [**Qui67**] Quillen introduced the notion of a model category, but then almost right away only concerned himself with the slightly more restricted but also more useful *closed model categories* which in [**Qui69**] he characterized by five simple axioms. However since then it has become clear that it is more convenient to restrict the definition even further by

(i) strengthening his *limit axiom*, which requires the existence of all *finite* colimits and limits, by requiring the existence of all *small* colimits and limits, and

(ii) strengthening his *factorization axiom* by requiring the therein mentioned factorizations to be *functorial*,

and we will therefore use the term **model category** for a closed model category which satisfies these stronger versions of Quillen's limit and factorization axioms.

2.2. Categories. We will use the term **locally small category** for what one usually calls a category, i.e. a category with *small* hom-sets, and reserve the term **category** for a more general notion which ensures that

(i) for every two categories C and D, the functors $C \to D$ and the natural transformations between them also form a category (which is usually denoted by D^C), and

(ii) for every category C and subcategory $W \subset C$, there exists a **localization** of C with respect to W, i.e. a category $C[W^{-1}]$ together with a **localization functor** $\gamma\colon C \to C[W^{-1}]$ which is 1-1 and onto on objects and sends the maps of W to isomorphisms in $C[W^{-1}]$, with the universal property that, for every category B and functor $b\colon C \to B$ which sends the maps of W to isomorphisms in B, there is a unique functor $b_\gamma\colon C[W^{-1}] \to B$ such that $b_\gamma \gamma = b$.

It is also convenient to consider, *for use in this chapter only*, the notion of

2.3. Categories with weak equivalences. By a **category with weak equivalences** or **we-category** we will mean a category C with a single distinguished class W of maps (called **weak equivalences**) such that

(i) W contains all the isomorphisms, and

(ii) W has Quillen's **two out of three property** that, for every pair of maps $f, g \in C$ such that gf exists, if *two* of f, g and gf are in W, then so is the *third*

which readily implies that

(iii) W is a subcategory of C.

For such a we-category C one then can

(iv) define the **homotopy category** $\operatorname{Ho} C$ of C as the category obtained from C by "formally inverting" the weak equivalences, i.e. the category with the same objects as C in which, for every pair of objects $X, Y \in C$, the hom-set $\operatorname{Ho} C(X, Y)$ consists of the equivalence classes of zigzags in C from X to Y in which the backward maps are weak equivalences, by the weakest equivalence relation which puts two zigzags in the same class when one can be obtained from the other by

(a) omitting an identity map,

(b) replacing two adjacent maps which go in the same direction by their composition, or

(c) omitting two adjacent maps when they are the same but go in opposite directions,

note that this homotopy category $\operatorname{Ho} C$, together with the functor $\gamma\colon C \to \operatorname{Ho} C$ which is the identity on the objects and which sends a map $c\colon X \to Y \in C$ to the class containing the zigzag which consists of the map c only, is a *localization* of C with respect to its subcategory of weak equivalences (2.2(ii)) and call C **saturated** whenever a map in C is a weak equivalence iff its image under the localization functor $C \to \operatorname{Ho} C$ is an isomorphism in $\operatorname{Ho} C$, and

(v) given two we-categories C and D with localization functors $\gamma\colon C \to \operatorname{Ho} C$ and $\gamma'\colon D \to \operatorname{Ho} D$, and a functor $g\colon C \to D$ (which is not required to preserve weak equivalences), define a **total left** (resp. **right**) **derived functor** of g as a pair (n, c) consisting of a functor $n\colon \operatorname{Ho} C \to \operatorname{Ho} D$ and a natural transformation

$$c\colon n\gamma \longrightarrow \gamma'g \qquad (\text{resp. } c\colon \gamma'g \longrightarrow n\gamma)$$

which is a terminal (resp. an initial) object in the category of all such pairs.

3. Problems involving the homotopy category

We now discuss four problems that came up when we started to review some of the classical results involving the homotopy categories of model categories. The first of these involved the notion of

3.1. Saturation. Our problem with saturation was that, while this notion was defined in terms of only the weak equivalences (2.3(iv)), the classical proof of the fact that

(i) every model category is a saturated we-category (2.3)

heavily involved the cofibrations and the fibrations and was very ad-hoc. We felt that it should have been possible to describe "reasonable" conditions on an arbitrary we-category C such that

(ii) these conditions imply the saturation of C

and then show that

(iii) these conditions held for every model category.

However at that time we did not have the faintest idea what such conditions might look like.

A similar problem concerned

3.2. The hom-sets. The usual description, for a cofibrant object X_c and a fibrant object Y_f in a model category M, of the hom-set $\operatorname{Ho} M(X_c, Y_f)$ (2.3(iv)) as a quotient of the hom-set $M(X_c, Y_f)$ by a homotopy relation, readily implied that

(i) for every pair of objects $X, Y \in M$, the elements of $\operatorname{Ho} M(X, Y)$ are in a natural 1-1 correspondence with the equivalence classes of the zigzags in M of the form

$$X \longleftarrow \cdot \longrightarrow \cdot \longleftarrow Y$$

in which the two backward maps are weak equivalences, where two such zigzags are in the same class iff they are the top row and the bottom row in a commutative diagram in M of the form

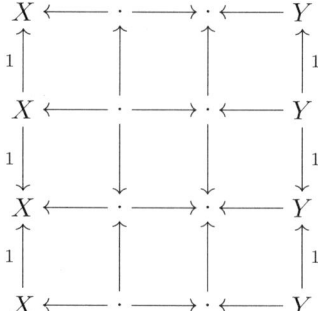

in which the backward maps (and hence all the vertical maps) are weak equivalences.

This suggested a question similar to the one raised in 3.1, but in this case we knew the answer, which was provided by the results of [**DK80a**] which implied that

    (ii) a sufficient condition on a we-category C (2.3) in order that (i) holds with C instead of M, is that C admits a **3-arrow calculus**, i.e. that there exist subcategories U and V of the category W of the weak equivalences in C such that, in a *functorial* manner

    (iii) for every zigzag $A' \xleftarrow{u} A \xrightarrow{f} B$ in C with $u \in U$, there exists a zigzag $A' \xrightarrow{f'} B' \xleftarrow{u'} B$ in C with $u' \in U$ such that $u'f = f'u$ and u' is an isomorphism whenever u is,

    (iv) for every zigzag $X \xrightarrow{g} Y \xleftarrow{v} Y'$ in C which $v \in V$, there exists a zigzag $X \xleftarrow{v'} X' \xrightarrow{g'} Y'$ in C with $v' \in V$ such that $gv' = vg'$ and v' is an isomorphism whenever v is, and

    (v) every map $w \in W$ admits a factorization $w = vu$ with $u \in U$ and $v \in V$.

and the fact that, in view of the assumption made in (2.1(ii))

    (vi) every model category admits a 3-arrow calculus consisting of the subcategories of the trivial cofibrations and the trivial fibrations, i.e. the cofibrations and the fibrations which are also weak equivalences.

Next we considered

3.3. Certain subcategories of the homotopy category. A simpler problem arose from the fact that, if for a model category M one denotes its full subcategories spanned by the cofibrant and the fibrant objects respectively by M_c and M_f and their intersection by M_{cf}, then

    (i) the inclusions of M_{cf}, M_c, M_f and M in each other induce a commutative diagram

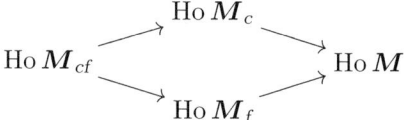

in which all the maps are equivalences of categories.

This made us, for an arbitrary we-category C (2.3), define

(ii) a **left** (resp. a **right**) **deformation** of C as a pair (r,s) consisting of a *weak equivalence preserving functor* $r\colon C \to C$ and a *natural weak equivalence*
$$s\colon r \longrightarrow 1_C \qquad (\text{resp. } s\colon 1_C \longrightarrow r)$$

and

(iii) a **left** (resp. a **right**) **deformation retract** of C as a *full* subcategory $C_0 \subset C$ for which there exists a left (resp. a right) deformation (r,s) of C into C_0, i.e. such that the functor $r\colon C \to C$ sends all of C into C_0,

and note that

(iv) for every left (resp. right) deformation retract $C_0 \subset C$, its inclusion $C_0 \to C$ induces an equivalence of categories $\operatorname{Ho} C_0 \to \operatorname{Ho} C$,

and that, in view of the assumption made in 2.1(ii)

(v) for every model category M there exist left and right deformations of M into M_c and M_f respectively (which are often referred to as **functorial cofibrant** and **fibrant replacements**,) so that M has M_c as a left deformation retract and M_f as a right deformation retract, and for similar reasons M_{cf} is a left deformation retract of M_f and a right deformation retract of M_c.

Our fourth and last problem involving the homotopy categories concerned the

3.4. Total derived functors of Quillen functors. Given a Quillen adjunction, i.e. an adjoint pair of functors $f\colon M \leftrightarrow N :f'$ between model categories of which the left adjoint f preserves cofibrations and trivial cofibrations (3.2(vi)) and the right adjoint f' preserves fibrations and trivial fibrations,

(i) the left adjoint f has total left derived functors (2.3(v)) and the right adjoint f' has total right derived functors, and

(ii) for every pair consisting of a total left derived functor (n,c) of f and a total right derived functor (n',c') of f', the given adjunction induces a derived adjunction $n\colon \operatorname{Ho} M \leftrightarrow \operatorname{Ho} N :n'$,

which raised the problem of isolating the "weak equivalence properties" of Quillen adjunctions which ensured the truth of these two statements.

To deal with this we

(iii) called a functor $f\colon C \to D$ (as in 2.3(v)) **left** (resp. **right**) **deformable** if f was a weak equivalence preserving functor on a left (resp. a right) deformation retract of C (3.3(iii)) or equivalently, if there existed a **left** (resp. a **right**) **f-deformation**, i.e. a left (resp. a right) deformation (r,s) of C (3.3(ii)) such that f preserved weak equivalences on the full subcategory of C spanned by the image of r,

and noted that:

(iv) if $f\colon C \to D$ was such a left (resp. right) deformable functor, then g had total left (resp. right) derived functors (2.3(v)) (as in that case, for every left (resp. right) f-deformation (r,s), the pair $(\operatorname{Ho} fr, \gamma'fs)$ (2.3(v)) turned out to be such a total derived functor),

and that

(v) for every pair of adjoint functors $f\colon \boldsymbol{C} \leftrightarrow \boldsymbol{D} :f'$ between we-categories (2.3) of which the left adjoint f was left deformable and the right adjoint f' was right deformable, its adjunction induced, for every pair consisting of a total left derived functor (n, c) of f and a total right derived functor (n', c') of f', a **derived adjunction** $n\colon \operatorname{Ho}\boldsymbol{C} \leftrightarrow \operatorname{Ho}\boldsymbol{D} :n'$.

That (i) and (ii) above then indeed were special cases of (iv) and (v) followed readily from 3.3(v) and the observation that

(vi) as, for every Quillen adjunction $f\colon \boldsymbol{M} \leftrightarrow \boldsymbol{N} :f'$ the functor f preserved weak equivalences between cofibrant objects and the functor f' preserved weak equivalences between fibrant ones, the functors f and f' were respectively left and right deformable.

4. Problem involving the homotopy colimit functors

Very different problems came up when we tried to deal with homotopy colimit functors on arbitrary model categories. To explain these we first recall the various definitions of

4.1. Homotopy colimit functors. For most authors, given a model category $\boldsymbol{M}$ and a small category $\boldsymbol{D}$,

(i) a *homotopy $\boldsymbol{D}$-colimit functor* on $\boldsymbol{M}$ is a pair (k, a) consisting of a functor $k\colon \boldsymbol{M}^{\boldsymbol{D}} \to \boldsymbol{M}$ which sends *objectwise* weak equivalences in $(\boldsymbol{M}_c)^{\boldsymbol{D}}$ (2.2(i) and 3.3) to weak equivalences in $\boldsymbol{M}$ and a natural transformation $a\colon k \to \operatorname{colim}^{\boldsymbol{D}}$ (where $\operatorname{colim}^{\boldsymbol{D}}$ denotes an arbitrary but fixed $\boldsymbol{D}$-colimit functor on $\boldsymbol{M}$, i.e. an arbitrary but fixed left adjoint of the constant diagram functor $\boldsymbol{M} \to \boldsymbol{M}^{\boldsymbol{D}}$), which is obtained (as in [**Hir03**]) by choosing a so-called framing of $\boldsymbol{M}$ and then applying the formulas of [**BK72**],

but for other authors

(ii) a *homotopy $\boldsymbol{D}$-colimit functor* on $\boldsymbol{M}$ is a pair (k', a') obtained from one of the above mentioned homotopy $\boldsymbol{D}$-colimit functors by (objectwise) pre-composition with a functorial cofibrant replacement (3.3(v)), as a result of which the functor $k'\colon \boldsymbol{M}^{\boldsymbol{D}} \to \boldsymbol{M}$ sends *all* objectwise weak equivalences of $\boldsymbol{M}^{\boldsymbol{D}}$ to weak equivalences in $\boldsymbol{M}$.

The first of these notions is the more practical one and is often used in calculations, but the second is more "homotopically correct" and was therefore of more interest to us, especially as we realized that

(iii) each of these homotopically correct homotopy $\boldsymbol{D}$-colimit functors on $\boldsymbol{M}$ was of the form $(\operatorname{colim}^{\boldsymbol{D}} r, \operatorname{colim}^{\boldsymbol{D}} s)$ for a special kind of left $\operatorname{colim}^{\boldsymbol{D}}$-deformation (3.4(iii))

and that moreover, if one considered these homotopy colimit functors as objects of the we-category (2.3)

$$\bigl(\operatorname{Fun}_w(\boldsymbol{M}^{\boldsymbol{D}}, \boldsymbol{M}) \downarrow \operatorname{colim}^{\boldsymbol{D}}\bigr)$$

which has as objects the pairs (k, a) consisting of an (objectwise) weak equivalence preserving functor $k\colon \boldsymbol{M}^{\boldsymbol{D}} \to \boldsymbol{M}$ and a natural transformation $a\colon k \to \operatorname{colim}^{\boldsymbol{D}}$ and as maps and weak equivalences $(k_1, a_1) \to (k_2, a_2)$ the natural transformations and natural weak equivalences $t\colon k_1 \to k_2$ such that $a_2 t = a_1$, then

(iv) any two of these homotopically correct homotopy $\boldsymbol{D}$-colimit functors on $\boldsymbol{M}$ were weakly equivalent (i.e. could be connected by a zigzag of weak equivalences) and in fact were so in a "homotopically unique" manner.

Now we can formulate

4.2. The problem. The last two comments (4.1(iii) and (iv)) raised the question whether, for every left $\operatorname{colim}^{\boldsymbol{D}}$-deformation (r', s') (for instance one that was special as in 4.1(iii) with respect to some other model structure on $\boldsymbol{M}$ with the *same* weak equivalences), the pair $(\operatorname{colim}^{\boldsymbol{D}} r', \operatorname{colim}^{\boldsymbol{D}} s')$ would be weakly equivalent to the pairs considered in 4.1(iii). We could not imagine that this was not the case and therefore wondered whether there existed, for every we-category $\boldsymbol{C}$ (2.3) on which there existed $\boldsymbol{D}$-colimit functors (4.1(i)) and for which we denoted by $\operatorname{colim}^{\boldsymbol{D}}$ an arbitrary but fixed such functor, we could define a class of objects in the we-category (4.1(iii))
$$\bigl(\operatorname{Fun}_{\mathrm{w}}(\boldsymbol{C}^{\boldsymbol{D}}, \boldsymbol{C}) \downarrow \operatorname{colim}^{\boldsymbol{D}}\bigr)$$
such that

(i) any two such objects were weakly equivalent in some homotopically unique manner, and
(ii) if $\boldsymbol{C}$ were the underlying we-category of a model category, then these objects included all the homotopically correct homotopy $\boldsymbol{D}$-colimit functors on this model category,

in which case all these objects deserved to be called homotopy $\boldsymbol{D}$-colimit functors on $\boldsymbol{C}$.

It seemed that such objects would have to be (objectwise) weak equivalence preserving functors $\boldsymbol{C}^{\boldsymbol{D}} \to \boldsymbol{C}$ over the functor $\operatorname{colim}^{\boldsymbol{D}}$ which "approached $\operatorname{colim}^{\boldsymbol{D}}$ as closely as possible from the left". They could however not be required to be terminal objects in the category $\bigl(\operatorname{Fun}_{\mathrm{w}}(\boldsymbol{C}^{\boldsymbol{D}}, \boldsymbol{C}) \downarrow \operatorname{colim}^{\boldsymbol{D}}\bigr)$, as in that case any two of them would be isomorphic, but could only be expected to be some kind of

4.3. Homotopically terminal objects. Given a we-category $\boldsymbol{Y}$ and an object $Y \in \boldsymbol{Y}$, we denoted by $1_{\boldsymbol{Y}} \colon \boldsymbol{Y} \to \boldsymbol{Y}$ the identity functor of $\boldsymbol{Y}$ and by $\operatorname{cst}_Y \colon \boldsymbol{Y} \to \boldsymbol{Y}$ the constant functor which sends all maps of $\boldsymbol{Y}$ to the identity map 1_Y of the object Y, and motivated by the fact that (see 38.1) Y would be a terminal object of $\boldsymbol{Y}$ *iff* there existed a natural transformation $f \colon 1_{\boldsymbol{Y}} \to \operatorname{cst}_Y$ such that the map $fY \colon Y \to Y$ was an *isomorphism*, we decided to call Y a **homotopically terminal object** of $\boldsymbol{Y}$ if there existed functors $F_0, F_1 \colon \boldsymbol{Y} \to \boldsymbol{Y}$ and a natural transformation $f \colon F_1 \to F_0$ such that

(i) F_0 was naturally weakly equivalent (i.e. could be connected by a zigzag of natural weak equivalences) to cst_Y,
(ii) F_1 was naturally weakly equivalent to $1_{\boldsymbol{Y}}$, and
(iii) the map $fY \colon F_1 Y \to F_0 Y$ was a *weak equivalence*.

Given a we-category $\boldsymbol{C}$ as in 4.2, the homotopically terminal objects of the we-category $\bigl(\operatorname{Fun}_{\mathrm{w}}(\boldsymbol{C}^{\boldsymbol{D}}, \boldsymbol{C}) \downarrow \operatorname{colim}^{\boldsymbol{D}}\bigr)$ then had the property mentioned in 4.2(ii), but in order to see whether they also satisfied 4.2(i) we first had to decide what exactly should be meant by

4.4. Homotopical uniqueness. Homotopical uniqueness had to be a kind of we-category version of the categorical notion of *uniqueness up to a (unique) isomorphism* (which we like to refer to as *categorical uniqueness*) of certain objects in a category Y, by which one meant that those objects (for instance because they had a common universal property) had the property that, for any two of them, Y_1 and Y_2, there was *exactly one* map $Y_1 \to Y_2 \in Y$ and that moreover this unique map was an *isomorphism*. Another way of saying that was that

(i) given a non-empty set of objects in a category Y, these objects were **categorically unique** or **canonically isomorphic** if the *full* subcategory $G \subset Y$ spanned by these objects *and all isomorphic ones* was **categorically contractible** in the sense that G was a non-empty groupoid in which there was exactly one isomorphism between any two objects, or equivalently that, for one (and hence every) object $G \in G$, the identity functor $1_G \colon G \to G$ and the constant functor $\mathrm{cst}_G \colon G \to G$ (4.3) were naturally isomorphic.

We therefore similarly said that

(ii) given a non-empty set of objects in a we-category Y, these objects were **homotopically unique** or **canonically weakly equivalent** if the full subcategory $G \subset Y$ spanned by these objects *and all weakly equivalent ones* (4.1(iv)) was **homotopically contractible** in the sense that, for one (and hence every) object $G \in G$, the functors $1_G \colon G \to G$ and $\mathrm{cst}_G \colon G \to G$ were naturally weakly equivalent (4.3(i)).

At this point we had hoped, in order to settle the question that we raised at the end of 4.2, that just as

(iii) in every category, the terminal objects, if they existed, would be canonically isomorphic (i),

it would also be true that

(iv) in every we-category, the homotopically terminal objects (4.3), if they existed, would be canonically weakly equivalent (ii).

However it turned out that this was *not* the case, unless one put some restrictions on the we-category considered, and after much trial and error we concluded that the minimal such restriction would be to require a property which was only slightly stronger than the "two out of three" property (2.3) and which, for want of a better name, we called

4.5. The two out of six property. We said that a we-category C had the **two out of six property** if

(i) for every three maps r, s and $t \in C$ for which the *two* compositions sr and ts were weak equivalences, the *four* maps r, s, t and tsr also were weak equivalences

and as (as we mentioned at the end of 4.4)

(ii) the presence of the two out of six property implied that the homotopically terminal objects (4.3), if they existed, were canonically weakly equivalent (4.4)

we could now finally answer the question which we raised at the end of 4.2 by defining, for every we-category as in 4.2 *which had the two out of six property*, and every small category D,

(iii) a **homotopy D-colimit functor** on C as a homotopically terminal object (4.3) of the we-category
$$\left(\operatorname{Fun}_w(C^D, C) \downarrow \operatorname{colim}^D\right)$$
(which clearly inherited the two out of six property from C)

and note that in view of (ii)

(iv) such homotopy D-colimit functors on C, if they existed, were homotopically unique

and that

(v) a sufficient condition for the existence of such functors was that the functor $\operatorname{colim}^D \colon C^D \to C$ be left deformable (3.4(iii)) as in that case, for every left colim^D-deformation (r, s), the pair $(\operatorname{colim}^D r, \operatorname{colim}^D s)$ was a homotopy D-colimit functor on C.

5. The emergence of the current monograph

Ever since the discovery of simplicial localizations [**DK80a, DK80b, DK80c**] we had wondered whether there might not exist some reasonable theory of we-categories (2.3) which would on the one hand (see 25.1) be a relative version of ordinary category theory and on the other hand provide a framework for homotopical algebra. These thoughts caused us, some ten years later, to ask the questions which we discussed in the preceding two sections and the answers we came up with suggested that there might indeed be such a theory, however not of we-categories, but of what we decided to call

5.1. Homotopical categories. We defined a **homotopical category** as a category C with a single distinguished class of maps (called **weak equivalences**) such that

(i) these weak equivalences had the two out of six property that (4.5(i)), for every three maps r, s and $t \in C$ for which the *two* compositions sr and ts were weak equivalences, the *four* maps r, s, t and tsr were also weak equivalences, and

(ii) for every object $C \in C$ its identity map $1_C \colon C \to C$ was also a weak equivalence

which readily implied, by considering the cases in which sr and ts are both identity maps or in which at least one of r, s and t is an identity map, that

(iii) all isomorphisms in C were weak equivalences, and C had the two out of three property, so that

(iv) C was a we-category (2.3)

However we still were not sure that this was the right setting until we realized that

(v) *every homotopical category which admits a 3-arrow calculus (3.2) is saturated (2.3(iv))*,

which, in view of 3.2(vi), finally provided an answer to the rather basic question which we raised in 3.1.

At this point we decided to give up on our original plan (see §1) and write instead the current monograph which consists of two essentially self contained parts, of which the first deals with *model categories* and the second with *homotopical categories*.

Part I which, apart from this introductory chapter, consists of the chapters II–IV, deals mainly with the model category results which we mentioned before (in §3 and §4), but does so, whenever possible, from the "homotopical" point of view which we develop in part II. More precisely, in the chapters II and III we introduce *model categories*, their *homotopy categories* and *Quillen functors* and discuss the results we mentioned in §3, while in chapter IV we not only prove the existence, on every model category, of *homotopy colimit* and *limit functors* (see §4), but also the considerably stronger result that (in an appropriate sense) every model category is *homotopically cocomplete* and *complete*.

Part II is concerned with *homotopical categories* only and consists of the four chapters V–VIII, of which the first is just a summary of the other three. These three chapters, of which we will give a preview in the next (and last) section of this introductory chapter, parallel the chapters II–IV and deal respectively with *homotopical categories* and their *homotopy categories*, *deformable functors* (3.4(iii)) and *homotopy colimit* and *limit functors* and more generally *homotopical colimit* and *limit functors* and the associated notions of *homotopical cocompleteness* and *completeness*.

6. A preview of part II

We end this chapter with a brief outline of the contents of part II, i.e. of the chapters VI–VIII.

6.1. Homotopical categories and homotopical functors. In the first part of chapter VI we

(i) motivate the somewhat unconventional use of the term *category* mentioned in 2.2,
(ii) define **homotopical categories** (5.1) and **homotopical functors** (i.e. weak equivalence preserving functors) between them, and
(iii) discuss a few immediate consequences of these definitions.

6.2. The homotopy category of a homotopical category. The middle part of chapter VI is concerned with the **homotopy category** Ho C of a homotopical category C (2.3(iv)) and in it we

(i) give an *alternate description* of Ho C as the category which has the same objects as C in which, for every pair of objects $X, Y \in C$, the hom-set Ho $C(X, Y)$ consists of the equivalence classes of zigzags in C from X to Y in which the backward maps are weak equivalences, with respect to the weakest equivalence relation which puts two such zigzags in the same class

(i)′ when one can be obtained from the other by omitting identity maps,
(i)″ when one can be obtained from the other by replacing adjacent maps which go in the same direction by their composition, and
(i)‴ when they are the top and the bottom row in a rectangular diagram in C of the form

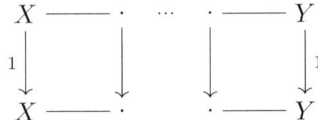

in which in every column the two horizontal maps go in the same direction and use this description to show that, if C admits a **3-arrow calculus** (3.2(ii)), then

  (ii) the hom-sets of Ho C admit the 3-*arrow description* given in 3.2(i),

which in turn implies that

  (iii) C has the rather useful property (see 6.5(iii) and 6.7 below) of being **saturated** (2.3(iv)).

We also note that every homotopical category C has a **Grothendieck enrichment** which is a category enriched over categories which has the same objects as C and in which, for every pair of objects $X, Y \in C$, the associated hom-category has as its objects all the zigzags in C from X to Y in which the backward maps are weak equivalences and show that this enriched category has the properties that

  (iv) replacement of each of its hom-categories by its set of *connected components* yields exactly the *homotopy category* Ho C, and
  (v) replacement of each of its hom-categories by its *nerve* yields a category enriched over simplicial sets which is weakly equivalent to the *(hammock) simplicial localization* of C of [**DK80a**].

6.3. Homotopically terminal and initial objects and Kan extensions. In the last part of chapter VI we

  (i) discuss in more detail the notions of **homotopically terminal** and **initial objects** (4.3) and **homotopical uniqueness** (4.4), and
  (ii) note that just as, given two (not necessarily homotopical) functors between homotopical categories
  $$p\colon R \longrightarrow P \quad \text{and} \quad q\colon R \longrightarrow Q$$
  one can define the right (or as we will say **terminal**) **Kan extensions** of q along p as the *terminal objects* of the homotopical category $(-p \downarrow q)$ which has as objects the pairs (r, e) consisting of a (not necessarily homotopical) functor $r\colon P \to Q$ and a natural transformation $e\colon rp \to q$, and as maps and weak equivalences $t\colon (r_1, e_1) \to (r_2, e_2)$ the natural transformations and natural weak equivalences $t\colon r_1 \to r_2$ such that $e_2(tp) = e_1$, one can define **homotopically terminal Kan extensions** of q along p as the *homotopically terminal objects* of its full subcategory $(-p \downarrow q)_\mathrm{w} \subset (-p \downarrow q)$ spanned by the objects (r, e) for which r is a *homotopical* (6.1) functor, and
  (iii) define **initial Kan extension** and **homotopically initial Kan extensions** in a dual manner.

6.4. Deformable functors. In chapter VII we investigate **left** and **right deformable** functors (3.4(iii)). These are not necessarily homotopical functors between homotopical categories which have homotopical meaning because they have (homotopically unique (4.4(ii))) *left* and *right approximations* respectively, where we

  (i) define, for a (not necessarily homotopical) functor $f\colon X \to Y$ between homotopical categories, a **left approximation** of f as a *homotopically* *terminal Kan extension* of f along $1_{\boldsymbol{X}}$ (6.3(ii)), i.e. a *homotopically terminal* object (4.3) of the "homotopical category of homotopical functors $\boldsymbol{X} \to \boldsymbol{Y}$ over f" (cf. 4.1(iii))

$$\left(\operatorname{Fun}_{\mathrm{w}}(\boldsymbol{X}, \boldsymbol{Y}) \downarrow f\right) .$$

They have the properties that

(ii) *such left approximations of f, if they exist, are homotopically unique* (4.4(ii) and 4.5(ii)),

(iii) a *sufficient* condition for their existence is that f be *left deformable* (3.4(iii)) as in that case *for every left f-deformation (r, s) (3.4(iii)), the pair (fr, fs) is a left approximation of f*,

(iv) *if f is left deformable, then, for every left approximation (k, a) of f, the pair $(\operatorname{Ho} k, \gamma' a)$ (where $\gamma' \colon \boldsymbol{Y} \to \operatorname{Ho} \boldsymbol{Y}$ denotes the localization functor of $\boldsymbol{Y}$ (2.2(ii))) is a total left derived functor of f* (2.3(v)), and

(v) for every composable pair of left deformable functors

$$f_1 \colon \boldsymbol{X} \longrightarrow \boldsymbol{Y} \quad \text{and} \quad f_2 \colon \boldsymbol{Y} \longrightarrow \boldsymbol{Z} ,$$

a *sufficient* condition in order that their composition $f_2 f_1 \colon \boldsymbol{X} \to \boldsymbol{Z}$ is also left deformable and every (appropriately defined) composition of their left approximations is a left approximation of their composition, is that the pair (f_1, f_2) is **left deformable** in the sense that

(vi) there exist left deformation retracts $\boldsymbol{X}_0 \subset \boldsymbol{X}$ and $\boldsymbol{Y}_0 \subset \boldsymbol{Y}$ (3.3(iii)) such that f_1 and $f_2 f_1$ are both homotopical (6.1) on $\boldsymbol{X}_0$ and f_2 is homotopical on $\boldsymbol{Y}_0$, and

(vi)' f_1 sends all of $\boldsymbol{X}_0$ into $\boldsymbol{Y}_0$

which turns out to be equivalent to requiring that

(vii) there exist left deformations (r_1, s_1) of $\boldsymbol{X}$ and (r_2, s_2) of $\boldsymbol{Y}$ (3.3(ii)) such that (r_1, s_1) is both a left f_1- and a left $f_2 f_1$-deformation (3.4(iii)) and (r_2, s_2) is a left f_2-deformation, and

(vii)' the natural transformation

$$f_2 s_2 f_1 r_1 \colon f_2 r_2 f_1 r_1 \longrightarrow f_2 f_1 r_1$$

is a natural weak equivalence.

Dual results of course hold for **right deformable functors**.

6.5. Deformable adjunctions. We also note in chapter VII that deformable functors often are part of a **deformable adjunction**, i.e. an adjunction $f \colon \boldsymbol{X} \leftrightarrow \boldsymbol{Y} \colon f'$ between homotopical categories of which the left adjoint is left deformable (3.4(iii)) and the right adjoint is right deformable.

Such deformable adjunctions have *derived adjunctions* (3.4(v)). However, as we noted that in dealing with these derived adjunctions we only used the total derived functors mentioned in 6.4(iv), it seemed to make more sense to describe derived adjunctions in terms of approximations instead. Instead of 3.4(v) we therefore

(i) show that every deformable adjunction $f \colon \boldsymbol{X} \leftrightarrow \boldsymbol{Y} \colon f'$ induces, for every pair consisting of a left approximation (k, a) of f and a right approximation (k', a') of f', an adjunction (called **derived adjunction**)

$$\operatorname{Ho} k \colon \operatorname{Ho} \boldsymbol{X} \longleftrightarrow \operatorname{Ho} \boldsymbol{Y} \colon \operatorname{Ho} k'$$

which is *natural* in (k, a) and (k', a').

We also note that

(ii) for every two composable deformable adjunctions
$$f_1 \colon \boldsymbol{X} \longleftrightarrow \boldsymbol{Y} : f_1' \qquad \text{and} \qquad f_2 \colon \boldsymbol{Y} \longleftrightarrow \boldsymbol{Z} : f_2'$$
a *sufficient* condition in order that their composition $f_2 f_1 \colon \boldsymbol{X} \leftrightarrow \boldsymbol{Z} : f_1' f_2'$ is also an adjoint pair of deformable functors and the compositions of their derived adjunctions are derived adjunctions of their composition is that the pairs (f_1, f_2) and (f_2', f_1') are respectively *left* and *right deformable* (6.4(v))

and that

(iii) for every two composable deformable adjunctions between *saturated* (2.3(iv)) homotopical categories
$$f_1 \colon \boldsymbol{X} \longleftrightarrow \boldsymbol{Y} : f_1' \qquad \text{and} \qquad f_2 \colon \boldsymbol{Y} \longleftrightarrow \boldsymbol{Z} : f_2'$$
for which the pairs (f_1, f_2) and (f_2', f_1') are **locally left** and **right deformable** in the sense that 6.4(vi) (or equivalently 6.4(vii)) holds but not necessarily 6.4(vi)′ (or 6.4(vii)′) *the pair (f_1, f_2) is actually left deformable* (6.4(v)) *iff the pair (f_2', f_1') is actually right deformable,*

This last result is rather useful as it often allows one, for a composable pair of left or right deformable functors between saturated homotopical categories, to verify the left deformability of this pair by only verifying its *local* left or right deformability in the sense of (iii).

6.6. A Quillen condition. At the end of chapter VII we note that, given a deformable adjunction (6.5) $f \colon \boldsymbol{X} \leftrightarrow \boldsymbol{Y} : f'$, a left deformation retract $\boldsymbol{X}_0 \subset \boldsymbol{X}$ on which f is homotopical (3.3(iii) and 6.1) and a right deformation retract $\boldsymbol{Y}_0 \subset \boldsymbol{Y}$ on which f' is homotopical

(i) for one (and hence every) pair consisting of a left approximation (k, a) of f and a right approximation (k', a') of f', the homotopical functor
$$k \colon \boldsymbol{X} \longrightarrow \boldsymbol{Y} \qquad \text{and} \qquad k' \colon \boldsymbol{Y} \longrightarrow \boldsymbol{X}$$
are **(inverse) homotopical equivalences of homotopical categories** in the sense that the two compositions $k'k$ and kk' are naturally weakly equivalent (4.3(i)) to the identity functors of $\boldsymbol{X}$ and $\boldsymbol{Y}$ respectively

iff the **Quillen condition** holds, i.e.

(ii) for every pair of objects $X_0 \in \boldsymbol{X}_0$ and $Y_0 \in \boldsymbol{Y}_0$, a map $f X_0 \to Y_0 \in \boldsymbol{Y}$ is a weak equivalence iff its adjunct $X_0 \to f' Y_0 \in \boldsymbol{X}$ is so.

6.7. Homotopy colimit and limit functors and homotopical ones. In the last chapter, chapter VIII, we note that, just as in 4.5(iii) we defined, for every small category $\boldsymbol{D}$, a homotopy $\boldsymbol{D}$-colimit functor on a homotopical category as a left approximation (6.4(i)) of an arbitrary but fixed $\boldsymbol{D}$-colimit functor $\operatorname{colim}^{\boldsymbol{D}} \colon \boldsymbol{X}^{\boldsymbol{D}} \to \boldsymbol{X}$ on $\boldsymbol{X}$, one can more generally, for every functor $u \colon \boldsymbol{A} \to \boldsymbol{B}$ between *not* necessarily small categories, define a **homotopy u-colimit functor** on $\boldsymbol{X}$ as a left approximation of an arbitrary but fixed u-colimit functor $\operatorname{colim}^{u} \colon \boldsymbol{X}^{\boldsymbol{A}} \to \boldsymbol{X}^{\boldsymbol{B}}$ on $\boldsymbol{X}$ (by which we mean an arbitrary but fixed left adjoint of the induced diagram functor $u^* \colon \boldsymbol{X}^{\boldsymbol{B}} \to \boldsymbol{X}^{\boldsymbol{A}}$). We then show that

(i) *such homotopy u-colimit functors on $\boldsymbol{X}$, if they exist, are homotopically unique* (4.4(ii)),

(ii) a *sufficient* condition for their existence is the existence of a *left deformable* (3.4(iii)) u-*colimit functor* on $\boldsymbol{X}$,

and that, for every composable pair of functors $u\colon \boldsymbol{A} \to \boldsymbol{B}$ and $v\colon \boldsymbol{B} \to \boldsymbol{D}$ between small categories

(iii) a *sufficient* condition in order that there exist homotopy u-colimit and v-colimit functors on $\boldsymbol{X}$ and that every (appropriately defined) composition of such a homotopy u-colimit functor with such a homotopy v-colimit functor is a homotopy vu-colimit functor on $\boldsymbol{X}$ is that there exist u-colimit and v-colimit functors on $\boldsymbol{X}$, and

(iii)' the pair $(\operatorname{colim}^u, \operatorname{colim}^v)$ is *left deformable* (6.4(v))

which is in particular the case if

(iii)'' $\boldsymbol{X}$ is *saturated* (2.3(iv)) and the pair $(\operatorname{colim}^u, \operatorname{colim}^v)$ is *locally left deformable* in the sense of 6.5(iii).

We also note that there is a corresponding notion of **homotopical cocompleteness** of a homotopical category $\boldsymbol{X}$ which is considerably stronger than the requirement that, for every small category $\boldsymbol{D}$, there exist homotopy $\boldsymbol{D}$-colimit functors on $\boldsymbol{X}$ and that

(iv) a *sufficient* condition for such homotopical cocompleteness of a homotopical category $\boldsymbol{X}$ is that $\boldsymbol{X}$ is *cocomplete* (which implies that, for every functor $u\colon \boldsymbol{A} \to \boldsymbol{B}$ between small categories, there exist u-colimit functors on $\boldsymbol{X}$) and that there exists, for every small category $\boldsymbol{D}$, a left deformation retract $(\boldsymbol{X}^{\boldsymbol{D}})_0 \subset \boldsymbol{X}^{\boldsymbol{D}}$ (3.3(iii)) such that, for every functor $u\colon \boldsymbol{A} \to \boldsymbol{B}$ between small categories, the u-colimit functor colim^u

(iv)' is *homotopical* on $(\boldsymbol{X}^{\boldsymbol{A}})_0$, and

(iv)'' sends *all* of $(\boldsymbol{X}^{\boldsymbol{A}})_0$ into $(\boldsymbol{X}^{\boldsymbol{B}})_0$,

which last restriction is superfluous if $\boldsymbol{X}$ is *saturated* (2.3(iv)).

We end the chapter with the observation that, while the above homotopy u-colimit functors are only defined for homotopical categories on which there exist u-colimit functors (and even then in addition require a choice of such a functor)

(v) there is also the notion of what we will call a **homotopical u-colimit functor** which, as we define such a functor as a *homotopically terminal Kan extension* (6.3(ii)) of the identity functor $1_{\boldsymbol{X}^{\boldsymbol{B}}}$ along the induced diagram functor $u^*\colon \boldsymbol{X}^{\boldsymbol{B}} \to \boldsymbol{X}^{\boldsymbol{A}}$, is defined for arbitrary homotopical categories and which "generalizes" the notion of a homotopy u-colimit functor in the sense that

(vi) if there exist u-colimit functors on $\boldsymbol{X}$, then

(vi)' there exist homotopical u-colimit functors on $\boldsymbol{X}$ *iff* there exist homotopy u-colimit functors on $\boldsymbol{X}$, and

(vi)'' in this case "composition with colim^u" yields a 1-1 *correspondence* between the homotopy u-colimit functors on $\boldsymbol{X}$ and the homotopical u-colimit functors on $\boldsymbol{X}$.

Moreover

(vii) there is a similar generalization of the notion of homotopical cocompleteness.

Of course dual results hold for **homotopy limit functors**, **homotopical completeness** and **homotopical limit functors**.

CHAPTER II

Model Categories and Their Homotopy Categories

7. Introduction

7.1. Summary. In this chapter we
  (i) introduce a notion of **model category** (i.e. a category M with three distinguished classes of maps, **weak equivalences**, **cofibrations** and **fibrations**, satisfying a few simple axioms which enable one to "do homotopy theory in M"), which is a slightly strengthened version of Quillen's notion of a closed model category ([**Hir03**], [**Hov99**]), and
  (ii) discuss the **homotopy category** of such a model category M, i.e. the category obtained from M by "formally inverting" its weak equivalences.

In more detail:

7.2. Model Categories. In [**Qui67**] Quillen introduced a notion of **model category** and called such a model category **closed** whenever any two of the three distinguished classes of maps (7.1) determined the third. Furthermore he characterized these closed model categories by five particularly nice axioms [**Qui69**] and noted that the requirement that a model category be closed is not a serious one. In fact he showed that every model category in which (as always seems to be the case) the class of the weak equivalences is closed under retracts, can be turned into a closed model category just by closing the other two classes under retracts.

The first and the fifth of Quillen's five axioms for a closed model category however are weaker than one would expect. The first axiom assumes the existence of finite colimits and limits, but not of arbitrary **small** ones and the fifth axiom assumes the existence of certain factorizations, but does not insist on their **functoriality**. This allows for the inclusion of various categories of finitely generated chain complexes, but also considerably complicates much of the theory.

In view of all this we therefore throughout this monograph use the term **model category** for a *closed model category in the sense of Quillen, which satisfies the above suggested stronger versions of this first and fifth axioms.*

7.3. The second and third axioms. Quillen's five axioms, with the first and the fifth axioms strengthened as indicated above (7.2) are perfectly adequate. However, for reasons which we will explain in 7.4 and 7.5 below, we prefer the *equivalent* (9.3) set of axioms obtained by
  (i) *strengthening* the second axiom which requires the weak equivalences to have the **two out of three** property (i.e. if f and g are maps such that gf is defined and two of f, g and gf are weak equivalences, then so is the third) by requiring them to have the **two out of six** property that, for every three maps r, s and t for which the *two* compositions sr and ts are defined and are weak equivalences, the *four* maps r, s, t and tsr

are also weak equivalences (which by restriction to the cases in which at least one of r, s and t is an identity map, implies that the two out of three property), and

(ii) compensating for this by *weakening* the third axiom which requires that the classes of the weak equivalences, the cofibrations and the fibrations are all three closed under retracts, by omitting this requirement for the class of the weak equivalences.

These changes are suggested by the following observations on

7.4. The role of the weak equivalences. A closer look at the notion of a model category reveals that the weak equivalences already determine its "homotopy theory", while the cofibrations and the fibrations provide additional structure which enables one to "do" homotopy theory, in the sense that, while many homotopy notions involved in doing homotopy theory can be defined in terms of the weak equivalences, the verification of many of their properties (e.g. their existence) requires the cofibrations and/or the fibrations. Moreover in dealing with model categories one often uses auxiliary categories which are not model categories, but which still have obvious "weak equivalences" induced by the weak equivalences of the model categories involved.

All this points to the desirability of having a better understanding of "categories with weak equivalences". But when one tries to develop such a "relative" category theory, and in particular one which allows for reasonable "relative" versions of such basic categorical notions as initial and terminal objects, colimit and limit functors, Kan extensions and universal properties, one realizes (see 4.3–4.5) that the weak equivalences should have some closure properties and that it is convenient to restrict oneself to what we will call

7.5. Homotopical categories. A **homotopical category** will be a category with a *single* distinguished class of maps, called **weak equivalences**, which

(i) has the *two out of six property* (7.3) and
(ii) contains all the *identity maps*

which (see 5.1) implies that it also

(iii) has the *two out of three* property (7.3), and
(iv) contains all the *isomorphisms*.

These homotopical categories will be investigated in Part II of this monograph where we define for them such notions as **homotopically initial** and **terminal objects**, **homotopical Kan extensions** and **homotopy** and **homotopical colimit** and **limit functors**, which all can be characterized by **homotopically universal properties** which ensure that such notions, if they exist, are **homotopically unique** in the sense that they form a **homotopically contractible category**, i.e. a non-empty homotopical category in which all maps are weak equivalences and for which its identity functor can be connected by a (finite) zigzag of natural transformations to a constant functor, i.e. a functor which sends all maps to a single identity map.

Model categories and the auxiliary categories referred to in 7.4 are all homotopical categories and the various *homotopy notions* one encounters when working with model categories are *homotopically unique* in the strong sense mentioned above. To emphasize this strong connection between model categories and homotopical categories we therefore prefer the modifications of the second and third axioms suggested in 7.3. Moreover, as the modified retract axiom no longer involves the weak equivalences, this modification of the axioms also emphasizes the fact that the *weak equivalences* play a *different* role than the *cofibrations and fibrations*.

7.6. The homotopy category. We also discuss the **homotopy category** of a model category $\boldsymbol{M}$, i.e. the category $\operatorname{Ho}\boldsymbol{M}$ obtained from $\boldsymbol{M}$ by "formally inverting" the weak equivalences and in particular note that

(i) *this homotopy category* $\operatorname{Ho}\boldsymbol{M}$ *is equivalent to the "classical homotopy category"* $\boldsymbol{M}_{cf}/\sim$ *of* $\boldsymbol{M}$,

which is the quotient of the full subcategory $\boldsymbol{M}_{cf} \subset \boldsymbol{M}$ spanned by the cofibrant fibrant objects by a homotopy relation $\sim$ on the maps of $\boldsymbol{M}$ which, on the maps of $\boldsymbol{M}_{cf}$ is an equivalence relation which is compatible with the composition. Moreover this category $\boldsymbol{M}_{cf}$ has the property that

(ii) *a map in* $\boldsymbol{M}_{cf}$ *is a weak equivalence iff its image in* $\operatorname{Ho}\boldsymbol{M}$ *is an isomorphism.*

The proofs of these results are rather long and technical and as several good versions have recently been published in [**DS95**], [**GJ99**], [**Hir03**] and [**Hov99**], we refer the reader to those sources.

7.7. Homotopical comments. From the results we just mentioned in 7.6(i) and (ii) one can deduce that

(i) *every model category* $\boldsymbol{M}$ *is* **saturated** *in the sense that a map in* $\boldsymbol{M}$ *is a weak equivalence iff its image in the homotopy category* $\operatorname{Ho}\boldsymbol{M}$ *is an isomorphism, and*

(ii) *for every pair of objects* X *and* Y *in a model category* $\boldsymbol{M}$, *the set of the maps* $X \to Y \in \operatorname{Ho}\boldsymbol{M}$ *can be considered as the set of the equivalence classes of the zigzags in* $\boldsymbol{M}$ *of the form*

$$X \longleftarrow \cdot \longrightarrow \cdot \longleftarrow Y$$

in which the backward maps are weak equivalences, where two such zigzags are in the same class iff they are the top row and the bottom row in a commutative diagram in $\boldsymbol{M}$ *of the form*

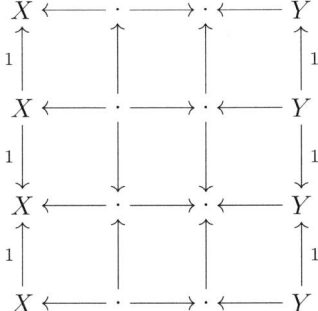

in which the backward maps (and hence the vertical maps) are weak equivalences.

As both these statements involve only the weak equivalences and not the cofibrations and the fibrations, it should be possible to find some reasonable properties of the weak equivalences which imply these two results and we will in fact show that both these results can be deduced from the observation that

(iii) M is a *homotopical category* (7.5), and
(iv) this homotopical category admits a so-called *3-arrow calculus*.

7.8. Organization of the chapter. There are four more sections. The first of these (§8) is devoted to some *categorical* and *homotopical preliminaries*. The next two sections deal with *model categories* (§9) and their *homotopy categories* (§10), while the last section (§11) elaborates on the homotopical comments of 7.7.

8. Categorical and homotopical preliminaries

In this section we

(i) indicate what exactly we will mean by *small categories*, *locally small categories* and just plain *categories*, and
(ii) recall from chapter VI (in part II of this monograph) some homotopical notions and results which will be needed in this chapter.

We thus start with a brief discussion of

8.1. Categories, locally small categories and small categories. In order to avoid set theoretical difficulties one often defines categories in terms of a *universe*, i.e. (see §32)

(i) one assumes that *every set is an element of some universe*, where one defines a **universe** as a set $\mathcal{U}$ of sets (called $\mathcal{U}$-**sets**) satisfying a few simple axioms which imply that $\mathcal{U}$ is closed under the usual operations of set theory and that every $\mathcal{U}$-set is a *subset* of $\mathcal{U}$, but that the set $\mathcal{U}$ itself and many of its subsets are *not* $\mathcal{U}$-sets, and
(ii) one then defines a $\mathcal{U}$-**category** as a category of which the hom-sets are $\mathcal{U}$-sets and the set of objects is a *subset* of $\mathcal{U}$ and calls such a $\mathcal{U}$-category **small** if the set of its objects is actually a $\mathcal{U}$-*set*.

It turns out that the notion of a small $\mathcal{U}$-category is indeed a convenient one, in the sense that any "reasonable" operation, when applied to small $\mathcal{U}$-categories yields again a small $\mathcal{U}$-category, but that the notion of a $\mathcal{U}$-category is not. For instance one readily verifies that

(iii) for every two *small* $\mathcal{U}$-categories C and D, the **diagram** or **functor category** $\text{Fun}(C, D)$ or D^C (which has as objects the functors $C \to D$ and as maps the natural transformations between them) is also a *small* $\mathcal{U}$-category, and that
(iv) for every *small* $\mathcal{U}$-category C and subcategory $W \subset C$, there exists a **localization** of C with respect to W, i.e. a *small* $\mathcal{U}$-category $C[W^{-1}]$, together with an often suppressed **localization functor** $\gamma \colon C \to C[W^{-1}]$ which is 1-1 and onto on objects, sends the maps of W to isomorphisms in $C[W^{-1}]$ and has the *universal property* that, for every category B and functor $b \colon C \to B$ which sends the maps of W to isomorphisms in B, there is a *unique* functor $b_\gamma \colon C[W^{-1}] \to B$ such that $b_\gamma \gamma = b$.

However both these statements are no longer true if one drops the requirement that the $\mathcal{U}$-categories involved are *small*. Still one cannot avoid the notion of $\mathcal{U}$-category as many of the categories one is interested in are not small and a way to get around this problem is to note that

(v) for every universe $\mathcal{U}$, there exists a unique **successor universe** $\mathcal{U}^+$, i.e. the unique smallest universe $\mathcal{U}^+$ such that the set $\mathcal{U}$ is a $\mathcal{U}^+$-set.

This implies that *every $\mathcal{U}$-category is a small $\mathcal{U}^+$-category* so that any "reasonable" operation when applied to $\mathcal{U}$-categories yields a small $\mathcal{U}^+$-category.

In view of all this **we choose an arbitrary but fixed universe $\mathcal{U}$** and use the term

> **small category**   for *small $\mathcal{U}$-category*, and
> **locally small category**   for *$\mathcal{U}$-category*

(instead of the customary use of the term *category* for *$\mathcal{U}$-category*) and reserve, **unless the context clearly indicates otherwise**, the term

> **category**   for *small $\mathcal{U}^+$-category*.

Similarly we will usually use the term

> **set**   for *$\mathcal{U}^+$-set*, and
> **small set**   for *$\mathcal{U}$-set*.

We then define as follows

8.2. Homotopical categories. A **homotopical category** will be a category $\boldsymbol{C}$ (8.1) with a distinguished set $\boldsymbol{W}$ (8.1) of maps (called **weak equivalences**) such that

(i) $\boldsymbol{W}$ contains all the *identity maps* of $\boldsymbol{C}$, and
(ii) $\boldsymbol{W}$ has the **two out of six property** that, for every three maps r, s and $t \in \boldsymbol{C}$ for which the *two* composition sr and ts exist and are in $\boldsymbol{W}$, the *four* maps r, s, t and tsr are also in $\boldsymbol{W}$

which one readily verifies (by assuming that both sr and ts are identity maps or that at least one of r, s and t is an identity map) is equivalent to requiring that

(i)′ $\boldsymbol{W}$ contains all the *identity maps* of $\boldsymbol{C}$,
(ii)′ $\boldsymbol{W}$ has the **weak invertibility property** that every map $s \in \boldsymbol{C}$ for which there exist maps r and $t \in \boldsymbol{C}$ such that the compositions sr and ts exist and are in $\boldsymbol{W}$, is itself in $\boldsymbol{W}$ (which, together with (i)′, implies that all *isomorphisms* of $\boldsymbol{C}$ are in $\boldsymbol{W}$), and
(iii)′ $\boldsymbol{W}$ has the **two out of three property** that, for every two maps f and $g \in \boldsymbol{C}$ for which gf exists and two of f, g and gf are in $\boldsymbol{W}$, so is the third (which implies that $\boldsymbol{W}$ is actually a *subcategory* of $\boldsymbol{C}$)

and two objects of $\boldsymbol{C}$ will be called **weakly equivalent** whenever they can be connected by a zigzag of weak equivalences. Furthermore we will refer to the category $\boldsymbol{W}$ as the **homotopical structure** of $\boldsymbol{C}$ and consider every subcategory $\boldsymbol{C}' \subset \boldsymbol{C}$ as a homotopical category with $\boldsymbol{W} \cap \boldsymbol{C}'$ as its homotopical structure.

There is of course the corresponding notion of

8.3. Homotopical functors. Given two homotopical categories C and D, a **homotopical functor** $C \to D$ will be a functor $C \to D$ which *preserves weak equivalences*.

One can also consider not necessarily homotopical functors $C \to D$ and call a natural transformation between two such functors a **natural weak equivalence** if it sends the objects of C to weak equivalences in D. The homotopical structure on D (8.2) then gives rise to a homotopical structure on the **diagram** or **functor category** (8.1(iii))

$$\operatorname{Fun}(C, D) \quad \text{or} \quad D^C$$

in which the weak equivalences are the natural weak equivalences, as well as (8.2) on its *full subcategory spanned by the homotopical functors* which we will denote by

$$\operatorname{Fun}_w(C, D) \quad \text{or} \quad (D^C)_w \ .$$

Furthermore, given two homotopical categories C and D, we

(i) call two homotopical functors $C \to D$ **naturally weakly equivalent** if they can be connected by a zigzag of natural weak equivalences, and

(ii) call a homotopical functor $f\colon C \to D$ a **homotopical equivalence of homotopical categories** if there exists a homotopical functor $f'\colon D \to C$ (called a **homotopical inverse** of f) such that the compositions $f'f$ and ff' are naturally weakly equivalent to the identity functors 1_C and 1_D respectively

and given a homotopical category C and a *full* subcategory $C_0 \subset C$, we

(iii) call C_0 a **left** (or a **right**) **deformation retract** of C if there exists a **left** (or a **right**) **deformation** of C into C_0, i.e. a pair (r, s) consisting of a *homotopical functor* $r\colon C \to C$ and a *natural weak equivalence*

$$s\colon r \longrightarrow 1_C \qquad (\text{or } s\colon 1_C \longrightarrow r)$$

such that $rC \in C_0$ for every object $C \in C$.

We end with discussing

8.4. The homotopy category of a homotopical category. Given a homotopical category C, its **homotopy category** $\operatorname{Ho} C$ will be the category obtained from C by "formally inverting" the week equivalences, i.e. the category with the same objects as C in which, for every pair of objects $X, Y \in C$, the hom-set $\operatorname{Ho} C(X, Y)$ consists of the equivalence classes of zigzags in C from X to Y in which the backward maps are weak equivalences, by the weakest equivalence relation which puts two zigzags in the same class when one can be obtained from the other by

(a) omitting an identity map,

(b) replacing two adjacent maps which go in the same direction by their composition, or

(c) omitting two adjacent maps when they are the same but go in opposite directions,

and this homotopy category $\operatorname{Ho} C$, together with the functor $\gamma\colon C \to \operatorname{Ho} C$ (called **localization functor**) which is the identity on the objects and which sends a map $c\colon X \to Y \in C$ to the class containing the zigzag which consists of the map c only, is a *localization* of C with respect to its subcategory of weak equivalences (8.1(iv)).

In view of the terminological assumptions make in 8.1, such homotopy categories always exist (see 33.8), although the homotopy category of a locally small homotopical category need not also be locally small.

As localizations were defined by means of a universal property (8.1(iv)), one readily verifies that

(i) *for every two homotopical categories C and D with localization functors $\gamma\colon C \to \operatorname{Ho} C$ and $\gamma'\colon D \to \operatorname{Ho} D$, and homotopical functor $f\colon C \to D$, there is a unique functor $\operatorname{Ho} f\colon \operatorname{Ho} C \to \operatorname{Ho} D$ such that $(\operatorname{Ho} f)\gamma = \gamma' f$, and*

(ii) *for every two such functors $f_1, f_2\colon C \to D$ and natural transformation (resp. natural weak equivalence) $h\colon f_1 \to f_2$, there is a unique natural transformation (resp. natural isomorphism) $\operatorname{Ho} h\colon \operatorname{Ho} f_1 \to \operatorname{Ho} f_2$ such that $(\operatorname{Ho} h)\gamma = \gamma' h$*

which implies that

(iii) *for every two homotopical functors $f_1, f_2\colon C \to D$ which are naturally weakly equivalent (8.3(i)), the functors $\operatorname{Ho} f_1, \operatorname{Ho} f_2\colon \operatorname{Ho} C \to \operatorname{Ho} D$ are naturally isomorphic, and*

(iv) *if a functor $f\colon C \to D$ is a homotopical equivalence of homotopical categories (8.3(ii)), then the functor $\operatorname{Ho} f\colon \operatorname{Ho} C \to \operatorname{Ho} D$ is an equivalence of categories.*

We end with mentioning that we will call a homotopical category C **saturated** when a map $c \in C$ is a weak equivalence *iff* the map $\gamma c \in \operatorname{Ho} C$ is an isomorphism, and noting (see 33.9(v)) that this definition readily implies that

(v) *if C is a saturated homotopical category, then so are, for every homotopical category D, the homotopical functor categories (8.3)*

$$\operatorname{Fun}(D, C) \quad \text{and} \quad \operatorname{Fun}_w(D, C) \ .$$

9. Model categories

We now

(i) define *model categories*, as mentioned in 7.2, as closed model categories in the sense of Quillen, but satisfying a strengthened version of his first and fifth axioms in which, as explained in 7.3, we modify his second and third axioms in a manner which does not affect the resulting notion of model category,

(ii) discuss a few immediate consequences of this definition, and

(iii) describe a few rather obvious examples.

We thus start with defining

9.1. Model categories. A **model category** will be a *locally small* category M (8.1) with a **model structure**, i.e. three sets (8.1) of maps, **weak equivalences**, **cofibrations** and **fibrations** (which are often denoted by $\xrightarrow{\sim}$, $\rightarrowtail$ and $\twoheadrightarrow$ respectively) satisfying the following five axioms:

MC1 Limit axiom: The category M is *cocomplete* and *complete*, i.e. there exists, for every small category D (8.1) a D-colimit and a D-limit functor $M^D \to M$ on X (i.e. a left and a right adjoint of the constant diagram functor $M \to M^D$).

MC2 Two out of six axiom: The weak equivalences have the *two out of six property*, i.e. if r, s and t are maps in $\boldsymbol{M}$ such that the *two* compositions sr and ts are defined and are weak equivalences, then the *four* maps r, s, t and tsr are also weak equivalences (8.2).

MC3 Retract axiom: The *cofibrations* and the *fibrations* are *closed under retracts*, i.e. if in a commutative diagram in $\boldsymbol{M}$ of the form

$$\begin{array}{ccccc} A & \xrightarrow{p} & X & \xrightarrow{q} & A \\ {\scriptstyle a}\downarrow & & {\scriptstyle x}\downarrow & & {\scriptstyle a}\downarrow \\ A' & \xrightarrow{p'} & X' & \xrightarrow{q'} & A' \end{array}$$

in which $qp = 1_A$ and $q'p' = 1_{A'}$, the map $x\colon X \to X'$ is a cofibration or a fibration, then so is the map $a\colon A \to A'$.

MC4 Lifting axiom:
(i) Every *cofibration* has the *left lifting property* with respect to every *trivial fibration* (i.e. a fibration which is also a weak equivalence and which therefore often will be indicated by $\xrightarrow{\sim}\!\!\!\!\to$), and
(ii) every *fibration* has the *right lifting property* with respect to every *trivial cofibration* (i.e. a cofibration which is also a weak equivalence and which therefore often will be indicated by $\xrightarrow{\sim}\!\!\!\!\hookrightarrow$),

by which one means that

(iii) for every commutative solid arrow diagram in $\boldsymbol{M}$

$$\begin{array}{ccc} A & \xrightarrow{f} & X \\ {\scriptstyle i}\downarrow & {\scriptstyle k}\nearrow & \downarrow{\scriptstyle p} \\ B & \xrightarrow{g} & Y \end{array}$$

in which i is a cofibration and p is a fibration, there exists a dotted arrow $k\colon B \to X$ such that $ki = f$ and $pk = g$ whenever at least one of i and p is a weak equivalence.

MC5 Factorization axiom: The maps $f \in \boldsymbol{M}$ admit *functorial factorizations*
(i) $f = qi$, where i is a cofibration and q is a trivial fibration, and
(ii) $f = pj$, where p is a fibration and j is a trivial cofibration.

One then has

9.2. Proposition. *Every model category $\boldsymbol{M}$ is a homotopical category* (8.2).

Proof. It suffices to prove that, for every object $X \in \boldsymbol{M}$, the identity map $1_X \colon X \to X \in \boldsymbol{M}$ is a weak equivalence and one does this by noting that the map 1_X admits (in view of MC5) a factorization $1_X = qi$ in which q is a trivial fibration, and then applying the two out of three property (8.2(iii)) to the equality $1_X q = q$.

Next we deal with the

9. MODEL CATEGORIES

9.3. Comparison with Quillen's closed model categories. To show that the above notion of a model category agrees with that of a closed model category in the sense of Quillen [**Qui69**] with his first and last axiom strengthened as in MC1 and MC5, one has to verify that

(i) the axioms MC1–5 imply that the weak equivalences are closed under retracts, and
(ii) the axioms MC1, MC3, MC4 and MC5, together with the closure of the weak equivalences under retract and the two out of three property (8.2(iii)) imply axiom MC2, i.e. the two out of six property (8.2(i)).

Such a verification of (i) will be given in 9.4 below, while (ii) follows from proposition 10.6 below (which states that every model category is saturated (8.4)) and the observation that its proof, for which we refer the reader to [**DS95**], [**GJ99**], [**Hir03**] or [**Hov99**], uses only the assumptions made in (ii).

It thus remains to prove

9.4. Proposition. *The weak equivalences in a model category are closed under retracts.*

Proof.

We have to show that, *if in the diagram of* MC3 *the map x is a weak equivalence then so is the map a*, and we do this by noting

(i) that, in view of MC5, the diagram of MC3 gives rise to a commutative diagram of the form (9.1)

$$\begin{array}{ccccccc}
A & \xrightarrow{p} & X & \xrightarrow{q} & A & \xrightarrow{\sim}_{j} & B \\
{\scriptstyle a'}\downarrow & & {\scriptstyle x'}\downarrow & & & {\scriptstyle r}\swarrow & \\
A'' & \xrightarrow{p''} & X'' & \xrightarrow{q''} & A'' & & \\
{\scriptstyle a''}\downarrow\sim & & {\scriptstyle x''}\downarrow\sim & & {\scriptstyle a''}\downarrow\sim & & \\
A' & \xrightarrow{p'} & X' & \xrightarrow{q'} & A' & &
\end{array}$$

in which $a''a' = a$, $x''x' = x$, $q''p'' = 1_{A''}$ and $rj = a'$ (and of course $qp = 1_A$ and $q'p' = 1_{A'}$),

(ii) that in view of the two out of three property and the assumption on the map x, the map x' is a weak equivalence,

(iii) that therefore (MC4) there exists a map $k\colon X'' \to B$ such that

$$rk = q'' \quad \text{and} \quad kx' = jq$$

which implies that

$$rkp'' = q''p'' = 1_{A''} \quad \text{and} \quad kp''a' = kx'p = jqp = j$$

(iv) and that, as (9.2) the map $1_{A''}$ is a weak equivalence, application of the two out of six property to the triple (a', kp'', r) yields that a' and r are weak equivalences, and so is therefore, in view of the two out of three property the map a.

Using this proposition we now prove the following useful

9.5. Closure properties of model categories. *Any two of the sets of the weak equivalences, the cofibrations and the fibrations in a model category (9.1) determine the third. In fact*
  (i) *a map is a cofibration iff it has the left lifting property with respect to all trivial fibrations,*
  (i)' *a map is a fibration iff it has the right lifting property with respect to all trivial cofibrations,*
  (ii) *a map is a trivial cofibration iff it has the left lifting property with respect to all fibrations,*
  (ii)' *a map is a trivial fibration iff it has the right lifting property with respect to all cofibrations, and*
  (iii) *a map is a weak equivalence iff it admits a factorization into a trivial cofibration followed by a trivial fibration.*

9.6. Corollary.
  (i) *Every pushout of a cofibration or a trivial cofibration is again a cofibration or a trivial cofibration and the composition of two cofibrations or two trivial cofibrations is again a cofibration or a trivial cofibration, and*
  (ii) *every pullback of a fibration or a trivial fibration is again a fibration or a trivial fibration and the composition of two fibrations or trivial fibrations is again a fibration or a trivial fibration.*

Proof of 9.5. Let f be a map which has the left lifting property with respect to all fibrations and (MC5) factor f into a trivial cofibration j followed by a fibration p. The lifting axiom (MC4) then yields a commutative diagram of the form

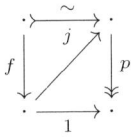

This readily implies that f is a retract of j and hence (MC3 and 9.4) a trivial cofibration, proving (ii).

The proofs of (i), (i)' and (ii)' are similar, while (iii) is a consequence of the factorization axiom (MC5) and the two out of three property (MC2).

We end with a few rather straightforward

9.7. Examples.
  (i) If $\boldsymbol{M}$ is a model category, then so is its *opposite* $\boldsymbol{M}^{\mathrm{op}}$, with the "opposites" of the weak equivalences as weak equivalences, the "opposites" of the fibrations as cofibrations and the "opposites" of the cofibrations as fibrations.
  (ii) For every object X in a model category $\boldsymbol{M}$, the *under category* $(X \downarrow \boldsymbol{M})$ and the *over category* $(\boldsymbol{M} \downarrow X)$ inherit from $\boldsymbol{M}$ a model structure in which a map is a weak equivalence, a cofibration or a fibration iff its image in $\boldsymbol{M}$ under the forgetful functor is so.

(iii) Every *product* of model categories, indexed by a small set (8.1), admits a model structure in which a map is a weak equivalence, a cofibration or a fibration iff all its projections onto the given model categories are so.

(iv) Every cocomplete and complete category admits a (unique) **minimal model structure** in which the weak equivalences are the *isomorphisms* and all maps are cofibrations as well as fibrations.

(v) Every cocomplete and complete category also admits **maximal model structures**, i.e., model structures in which *all maps* are weak equivalences. In one such model structure all maps are cofibrations, while the fibrations are the isomorphisms or dually one could require all maps to be fibrations and take for the cofibrations the isomorphisms. But there might be other maximal model structures as well. For instance in the category of small sets (8.1) one could take as cofibrations the monomorphisms and as fibrations the epimorphisms, or conversely as cofibrations the epimorphisms and as fibrations the monomorphisms.

10. The homotopy category

With every model category M one can associate its **homotopy category** $\operatorname{Ho} M$ (i.e. the category obtained from M be "formally inverting" the weak equivalences) and our main aim in this section is to obtain a more explicit description of this homotopy category by noting that

(i) the homotopy category $\operatorname{Ho} M$ is *equivalent* to the homotopy category $\operatorname{Ho} M_{cf}$ of the full subcategory $M_{cf} \subset M$ spanned by the **cofibrant fibrant objects**, and

(ii) the homotopy category $\operatorname{Ho} M_{cf}$ is canonically isomorphic to the quotient $M_{cf}/\sim$ of M_{cf} by a **homotopy relation** $\sim$.

These results imply that

(iii) the homotopy category $\operatorname{Ho} M$ is, just like the category M itself, *locally small*, and

(iv) the maps in $\operatorname{Ho} M$ can be considered as equivalence classes of zigzags in M of the form

in which the backward maps are weak equivalences, by an equivalence relation which can be formulated in terms of the weak equivalences.

We also note that

(v) a map in M_{cf} is a weak equivalence *iff* its image in $M_{cf}/\sim$ is an isomorphism which implies that

(vi) every model category M is *saturated*, i.e. (8.4) a map in M is a weak equivalence iff its image in $\operatorname{Ho} M$ is an isomorphism.

We start with recalling from 8.4

10.1. The homotopy category. Given a model category M (9.1), its **homotopy category** $\operatorname{Ho} M$ is the homotopy category (8.4) of its underlying homotopical category (9.2), i.e. the category with the same objects as M in which, for every pair of objects $X, Y \in M$, the hom-set $\operatorname{Ho} M(X, Y)$ consists of equivalence classes of zigzags in M from X to Y in which the backward maps are weak equivalences. It comes with a **localization functor** $\gamma \colon M \to \operatorname{Ho} M$ which is the identity on the objects and which sends a map $f\colon X \to Y \in \boldsymbol{M}$ to the class containing the zigzag which consists of the map f only.

In order to obtain a more explicit description of this homotopy category of $\boldsymbol{M}$ one first considers the full subcategories of $\boldsymbol{M}$ spanned by its

10.2. Cofibrant and fibrant objects. Given a model category $\boldsymbol{M}$, an object $X \in \boldsymbol{M}$ will be called **cofibrant** if the unique map to X from an initial object of $\boldsymbol{M}$ (which exists in view of the limit axiom (9.1)) is a cofibration, and dually an object $Y \in \boldsymbol{M}$ will be called **fibrant** if the unique map from Y to a terminal object is a fibration. Clearly (9.5)
  (i) *if $X \in \boldsymbol{M}$ is cofibrant and $f\colon X \to Y \in \boldsymbol{M}$ is a cofibration, then Y is cofibrant, and dually*
  (ii) *if $Y \in \boldsymbol{M}$ is fibrant and $f\colon X \to Y \in \boldsymbol{M}$ is a fibration, then X is fibrant.*

Let $\boldsymbol{M}_c$, $\boldsymbol{M}_f$ and $\boldsymbol{M}_{cf} = \boldsymbol{M}_c \cap \boldsymbol{M}_f$ denote the full subcategories of $\boldsymbol{M}$ spanned respectively by the *cofibrant* objects, the *fibrant* objects and the **cofibrant fibrant** ones, i.e. the objects which are both cofibrant and fibrant. Then the inclusions of $\boldsymbol{M}_{cf}$, $\boldsymbol{M}_c$, $\boldsymbol{M}_f$ and $\boldsymbol{M}$ in each other induce (8.4) functors between their homotopy categories Ho $\boldsymbol{M}_{cf}$, Ho $\boldsymbol{M}_c$, Ho $\boldsymbol{M}_f$ and Ho $\boldsymbol{M}$ and (i) and (ii) above together with the factorization axiom (9.1) readily imply

10.3. Proposition. *Let $\boldsymbol{M}$ be a model category (9.1). Then*
  (i) *$\boldsymbol{M}_c$ (10.2) is a left deformation retract of $\boldsymbol{M}$ (i.e. (8.3(iii)) $\boldsymbol{M}_c$ is a full subcategory of $\boldsymbol{M}$ for which there exists a pair (r,s) consisting of a weak equivalence preserving functor $r\colon \boldsymbol{M} \to \boldsymbol{M}_c$ and a natural weak equivalence (8.3) $s\colon r \to 1_{\boldsymbol{M}}$), and dually*
  (ii) *$\boldsymbol{M}_f$ is a right deformation retract of $\boldsymbol{M}$, while*
  (iii) *$\boldsymbol{M}_{cf}$ is a left deformation retract of $\boldsymbol{M}_f$ and a right deformation retract of $\boldsymbol{M}_c$.*

10.4. Corollary. *The inclusions of $\boldsymbol{M}_{cf}$, $\boldsymbol{M}_c$, $\boldsymbol{M}_f$ and $\boldsymbol{M}$ in each other induce (8.3(iii), 8.4(iv) and 10.3) a commutative diagram of functors*

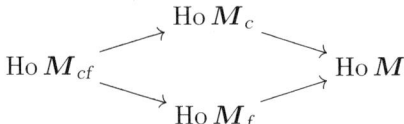

in which all functors are equivalences of categories.

To obtain a more explicit description of Ho $\boldsymbol{M}_{cf}$ (and hence of Ho $\boldsymbol{M}$) one next considers various

10.5. Homotopy relations. Given a model category $\boldsymbol{M}$ one can consider the following homotopy relations on the maps of $\boldsymbol{M}$:
  (i) **Left homotopic maps.** Two maps $f, g\colon B \to X \in \boldsymbol{M}$ will be called **left homotopic** if there exists a factorization
  $$B \amalg B \xrightarrowtail{i_0 \amalg i_1} B' \xrightarrow{\sim} B$$
  of the folding map $\nabla\colon B \amalg B \to B$ into a cofibration followed by a weak equivalence, together with a map $h\colon B' \to X \in \boldsymbol{M}$ such that $hi_0 = f$ and $hi_1 = g$.

(ii) **Right homotopic maps.** Dually two maps $f, g\colon B \to X \in \boldsymbol{M}$ will be called **right homotopic** if there exists a factorization
$$X \xrightarrow{\sim} X' \xrightarrow{p_0 \times p_1} X \times X$$
of the diagonal map $\Delta\colon X \to X \times X$ into a weak equivalence followed by a fibration, together with a map $k\colon B \to X' \in \boldsymbol{M}$ such that $p_0 k = f$ and $p_1 k = g$.

(iii) **Homotopic maps.** Two maps $f, g\colon B \to X \in \boldsymbol{M}$ will be called **homotopic** (denoted by $f \sim g$) if they are both left and right homotopic and a map $f\colon B \to X \in \boldsymbol{M}$ will be called a **homotopy equivalence** if there exists a map $f'\colon X \to B \in \boldsymbol{M}$ (called a **homotopy inverse** of f) such that $f'f \sim 1_B$ and $ff' \sim 1_X$.

The usefulness of these homotopy relations is due to the fact that not only clearly

(iv) *if two maps $f, g\colon B \to X \in \boldsymbol{M}$ are homotopic, left homotopic or right homotopic, then* (10.1) $\gamma f = \gamma g \in \operatorname{Ho} \boldsymbol{M}$,

but that, under suitable restrictions on B and X, these homotopy relations are equivalence relations for which the opposite of (iv) also holds. In fact one has the following

10.6. Properties of the homotopy relation. *Let $\boldsymbol{M}$ be a model category. Then, for every pair of objects $B \in \boldsymbol{M}_c$ and $X \in \boldsymbol{M}_f$ (10.2)*

(i) *the above (10.5) three homotopy relations agree on the set $\boldsymbol{M}(B, X)$ and are an equivalence relation on this set (which we will denote by $\sim$), and*

(ii) *this equivalence relation induces a map (10.5(iv))*
$$\boldsymbol{M}(B, X)/\sim \longrightarrow \operatorname{Ho} \boldsymbol{M}(B, X)$$
which is 1-1 and onto.

Moreover

(iii) *on the subcategory $\boldsymbol{M}_{cf} \subset \boldsymbol{M}$, this equivalence relation is compatible with the composition and thus induces a functor*
$$\boldsymbol{M}_{cf}/\sim \longrightarrow \operatorname{Ho} \boldsymbol{M}_{cf}$$
which is an isomorphism of categories, and

(iv) *a map in $\boldsymbol{M}_{cf}$ is a weak equivalence iff it is a homotopy equivalence (10.5(iii)) or equivalently (iii) iff its image in $\operatorname{Ho} \boldsymbol{M}_{cf}$ is an isomorphism.*

In view of 9.2, 10.3, 10.4 and 10.5 this result implies the following three corollaries.

10.7. Local smallness of the homotopy category. *For every model category $\boldsymbol{M}$,*

(i) *its homotopy category $\operatorname{Ho} \boldsymbol{M}$ (10.1) is equivalent to its "classical homotopy category" $\boldsymbol{M}_{cf}/\sim$ (10.6(iii)), and hence*

(ii) *its homotopy category $\operatorname{Ho} \boldsymbol{M}$ is, just like $\boldsymbol{M}$ itself (9.1), a locally small category (8.1).*

10.8. Saturation of the homotopy category. *Let $\boldsymbol{M}$ be a model category. Then $\boldsymbol{M}$ is a saturated homotopical category (8.4).*

10.9. 3-arrow description of the hom-sets of the homotopy category.
Let $\mathbf{M}$ be a model category and $\gamma\colon \mathbf{M} \to \operatorname{Ho} \mathbf{M}$ its localization functor. Then for every pair of objects $X, Y \in \mathbf{M}$,

(i) *the function which assigns to every zigzag in $\mathbf{M}$ of the form*

$$X \xleftarrow[\sim]{p} \cdot \xrightarrow{q} \cdot \xleftarrow[\sim]{r} Y$$

in which the backward maps are weak equivalences the map

$$(\gamma r)^{-1}(\gamma g)(\gamma p)^{-1}\colon X \longrightarrow Y \in \operatorname{Ho} \mathbf{M}$$

induces a 1-1 correspondence between the maps $X \to Y \in \operatorname{Ho} \mathbf{M}$ and the equivalence classes of the above zigzags, in which two such zigzags are in the same class whenever they are the top row and the bottom row in a commutative diagram in $\mathbf{M}$ of the form

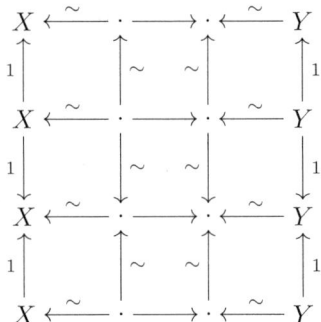

in which all the backward maps (and hence all the vertical maps) are weak equivalences, and

(ii) *for every map $f\colon X \to Y \in \mathbf{M}$, the equivalence class corresponding to the map $\gamma f\colon X \to Y \in \operatorname{Ho} \mathbf{M}$ is the class which contains the zigzag*

$$X \xleftarrow{1} X \xrightarrow{f} Y \xleftarrow{1} Y$$

It thus remains to prove proposition 10.6. However, as the proof of this proposition is rather long and technical and as good versions of it have recently been published in [**DS95**], [**GJ99**], [**Hir03**] and [**Hov99**], we refer the interested reader to those sources.

11. Homotopical comments

We ended the preceding section with two corollaries 10.8 and 10.9 in which we established two properties of model categories of which the formulation involved *only* the weak equivalences and *not* the cofibrations and the fibrations, and our aim in this section is to point out that there is a relatively simple property of the weak equivalences in a model category which implies these two properties.

To do this we start with recalling from 36.1 the notion of

11. HOMOTOPICAL COMMENTS

11.1. 3-arrow calculi. Given a homotopical category C with category of weak equivalences W (8.2), C is said to admit a **3-arrow calculus** $\{U, V\}$ if there exist subcategories $U, V \subset W$ such that in a *functorial* manner

(i) for every zigzag $A' \xleftarrow{u} A \xrightarrow{f} B$ in C which $u \in U$, there exists a zigzag $A' \xrightarrow{f'} B' \xleftarrow{u'} B$ in C with $u' \in U$ such that
$$u'f = f'u \text{ and}$$
u' is an isomorphism whenever u is

(e.g. if C is closed under pushouts and every pushout of a map in U is again in U),

(ii) for every zigzag $X \xrightarrow{g} Y \xleftarrow{v} Y'$ in C with $v \in V$, there exists a zigzag $X \xleftarrow{v'} X' \xrightarrow{g'} Y$ in C with $v' \in V$ such that
$$gv' = vg' \text{ and}$$
v' is an isomorphism whenever v is

(e.g. if C is closed under pullbacks and every pullback of a map in V is again in V), and

(iii) every map $w \in W$ admits a factorization $w = vu$ with $u \in U$ and $v \in V$.

Note that (iii) implies that U *and* V *contain all the objects of* W and hence of C.

We then showed in 36.3

11.2. 3-arrow description of the hom-sets of the homotopy category. Let C be a homotopical category (8.2) which admits a 3-arrow calculus (11.1) and let $\gamma\colon C \to \mathrm{Ho}\,C$ be its localization functor (8.4). Then, for every pair of objects $X, Y \in C$,

(i) *the function which assigns to every zigzag in* C *of the form*
$$X \xleftarrow[\sim]{p} \cdot \xrightarrow{q} \cdot \xleftarrow[\sim]{r} Y$$
in which the backward maps are weak equivalences, the map
$$(\gamma r)^{-1}(\gamma q)(\gamma p)^{-1}\colon X \longrightarrow Y \in \mathrm{Ho}\,C$$
induces a 1-1 correspondence between the maps $X \to Y \in \mathrm{Ho}\,C$ *and the equivalence classes of the above zigzags, in which two such zigzags are in the same class whenever they are the top row and the bottom row in a commutative diagram in* C *of the form*

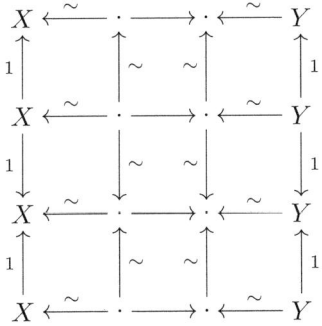

in which all the backward maps (and hence all the vertical maps) are weak equivalences, and

(ii) *for every map $f\colon X \to Y \in \boldsymbol{C}$, the equivalence class corresponding to the map $\gamma f\colon X \to Y \in \operatorname{Ho}\boldsymbol{C}$ is the class which contains the zigzag*

$$X \xleftarrow{\;\;1\;\;} X \xrightarrow{\;\;f\;\;} Y \xleftarrow{\;\;1\;\;} Y \ .$$

Furthermore, using this result, we prove in 36.4 that

11.3. 3-arrow calculi imply saturation. *Every homotopical category which admits a 3-arrow calculus is saturated (8.4).*

Corollaries 10.8 and 10.9 then follow immediately from 11.3 and 11.2 and the observation that 9.1, 9.2 and 9.6 readily imply the existence of

11.4. A 3-arrow calculus on model categories. *Every model category admits a 3-arrow calculus (11.1) consisting of the subcategories of the trivial cofibrations and the trivial fibrations.*

CHAPTER III

Quillen Functors

12. Introduction

12.1. Summary. We now consider a useful kind of "morphism between model categories" which is not, as one might expect, a functor which is compatible with the model structures, but an adjoint pair of functors, each of which is compatible with one half of the model structures involved. Neither of these two functors (which are called *left* and *right Quillen functors*) is required to be homotopical, i.e. (8.3) to preserve (all the) weak equivalences, but they still have "homotopical meaning": The left Quillen functor has *left approximations*, i.e. homotopical functors which, in a homotopical sense, are closest to it from the left, and which are *homotopically unique*, and the right Quillen functors have similar *right approximations*. Moreover these approximations have the convenient property that the functors between the homotopy categories induced by the left approximations are left adjoints of those induced by the right approximations.

In more detail:

12.2. Quillen functors. Given two model categories M and N, a **Quillen adjunction** between them will be an adjunction

$$f\colon M \longleftrightarrow N : f'$$

such that the left adjoint f (the **left Quillen functor**) preserves *cofibrations* and *trivial cofibrations* (9.1), while the right adjoint f' (the **right Quillen functor**) preserves *fibrations* and *trivial fibrations*. Thus neither functor is required to preserve weak equivalences. However they have the crucial property that

(i) *the left Quillen functor f preserves weak equivalences between cofibrant objects*, i.e., f restricts to a homotopical functor (8.3) on the left deformation retract M_c of M (10.2 and 10.3), and dually

(ii) *the right adjoint f' preserves weak equivalences between fibrant objects*, i.e. f' restricts to a homotopical functor on the right deformation retract M_f of M,

which ensures the existence of

12.3. Approximations. The usefulness of Quillen functors is that they have *approximations* which are *homotopically unique*. More precisely, a left Quillen functor $f\colon M \to N$ has **left approximations**, i.e. pairs (k, a) consisting of a *homotopical* functor $k\colon M \to N$ and a natural transformation $a\colon k \to f$, which in a homotopical sense are closest to f from the left. They are **homotopically unique**, in the sense that they form a homotopical category which is **homotopically contractible**, i.e. for which the (unique) functor to the terminal (homotopical) category [0] (which consists of only one object and its identity map) is a *homotopical equivalence of homotopical categories* (8.3).

Furthermore given two composable left Quillen functors $f_1\colon \boldsymbol{M} \to \boldsymbol{N}$ and $f_2\colon \boldsymbol{N} \to \boldsymbol{P}$, their composition $f_2 f_1 \colon \boldsymbol{M} \to \boldsymbol{P}$ is also a left Quillen functor and for every pair of left approximations (k_1, a_1) of f_1 and (k_2, a_2) of f_2, their **composition** $(k_2 k_1, a_2 a_1)$ in which $a_2 a_1$ is the diagonal of the commutative diagram

$$\begin{array}{ccc} k_2 k_1 & \xrightarrow{a_2} & f_2 k_1 \\ {\scriptstyle a_1}\downarrow & & \downarrow{\scriptstyle a_1} \\ k_2 f_1 & \xrightarrow{a_2} & f_2 f_1 \end{array}$$

is a left approximation of $f_2 f_1$.

Of course dually every right Quillen functor has right approximations with similar properties.

12.4. Derived adjunctions. A convenient property of the approximations of Quillen functors (12.3) is that, given a Quillen adjunction $f\colon \boldsymbol{M} \leftrightarrow \boldsymbol{N} \colon f'$, its adjunction induces, for every left approximation (k, a) of f and right approximation (k', a') of f', a **derived adjunction** between the homotopy categories of $\boldsymbol{M}$ and $\boldsymbol{N}$

$$\operatorname{Ho} k \colon \operatorname{Ho} \boldsymbol{M} \longleftrightarrow \operatorname{Ho} \boldsymbol{N} \colon \operatorname{Ho} k'$$

which is natural in (k, a) and (k', a') and which is *compatible with compositions* in the sense that, for every two composable Quillen adjunctions

$$f_1 \colon \boldsymbol{M} \longleftrightarrow \boldsymbol{N} \colon f'_1 \qquad \text{and} \qquad f_2 \colon \boldsymbol{N} \longleftrightarrow \boldsymbol{P} \colon f'_2$$

the compositions of their derived adjunctions are derived adjunctions of their composition.

We end with a brief discussion of Quillen adjunctions whose derived adjunctions are equivalences of categories and which are called

12.5. Quillen equivalences. It turns out that, given a Quillen adjunction $f\colon \boldsymbol{M} \leftrightarrow \boldsymbol{N} \colon f'$, the following six statements are equivalent:

(i) for one (and hence every) left approximation (k, a) of f, the functor $k\colon \boldsymbol{M} \to \boldsymbol{N}$ is a homotopical equivalence of homotopical categories (8.3),

(i)' for one (and hence every) right approximation (k', a') of f', the functor $k'\colon \boldsymbol{N} \to \boldsymbol{M}$ is a homotopical equivalence of homotopical categories,

(ii) for one (and hence every) pair consisting of a left approximation (k, a) of f and a right approximation (k', a') of f', the functors $k\colon \boldsymbol{M} \to \boldsymbol{N}$ and $k'\colon \boldsymbol{N} \to \boldsymbol{M}$ are homotopically inverse homotopical equivalences of homotopical categories (8.3),

(iii) for one (and hence every) left approximation (k, a) of f, the functor $\operatorname{Ho} k \colon \operatorname{Ho} \boldsymbol{M} \to \operatorname{Ho} \boldsymbol{N}$ is an equivalence of categories,

(iii)' for one (and hence every) right approximation (k', a') of f', the functor $\operatorname{Ho} k' \colon \operatorname{Ho} \boldsymbol{N} \to \operatorname{Ho} \boldsymbol{M}$ is an equivalence of categories, and

(iv) for one (and hence every) pair consisting of a left approximation (k, a) of f and a right approximation (k', a') of f', the functors $\operatorname{Ho} k \colon \operatorname{Ho} \boldsymbol{M} \to \operatorname{Ho} \boldsymbol{N}$ and $\operatorname{Ho} k' \colon \operatorname{Ho} \boldsymbol{N} \to \operatorname{Ho} \boldsymbol{M}$ are inverse equivalences of categories.

Thus a Quillen adjunction with these properties induces a kind of "equivalence of homotopy theories" and one therefore calls the left adjoint of such a Quillen adjunction a **left Quillen equivalence** and the right adjoint a **right Quillen equivalence**.

A useful *necessary and sufficient* condition in order that a Quillen adjunction $f \colon \boldsymbol{M} \leftrightarrow \boldsymbol{N} \colon f'$ be an adjunction of Quillen equivalences is the **Quillen condition** that, for every pair of objects $X \in \boldsymbol{M}_c$ and $Y \in \boldsymbol{N}_f$ (10.2), a map $X \to f'Y \in \boldsymbol{M}$ is a weak equivalence *iff* its adjunct $fX \to Y \in \boldsymbol{N}$ is so.

12.6. Homotopical comments. Most of the results of this chapter are special cases of more general results on homotopical categories, some of which we will need in chapter IV.

(i) Left and right Quillen functors (12.2) are special cases of **left** and **right deformable functors**, i.e. not necessarily homotopical functors between *homotopical* categories which are *homotopical on a left or a right deformation retract* (8.3) *of the domain category* and such left and right deformable functors have **left** or **right approximations** which are *homotopically unique* (12.3).

(ii) Given two composable left or right deformable functors $f_1 \colon \boldsymbol{X} \to \boldsymbol{Y}$ and $f_2 \colon \boldsymbol{Y} \to \boldsymbol{Z}$, their composition $f_2 f_1 \colon \boldsymbol{X} \to \boldsymbol{Z}$ need not again be left or right deformable, nor if it is, need the compositions of their left or right approximations be left or right approximations of their composition. However a *sufficient* condition for both these things to happen is that the pair (f_1, f_2) is what we will call **left** or **right deformable**.

(iii) Deformable functors often are part of a **deformable adjunction** (i.e. an adjunction $f \colon \boldsymbol{X} \leftrightarrow \boldsymbol{Y} \colon f'$ of which the left adjoint f is left deformable and the right adjoint f' is right deformable) and the adjunction of such a deformable adjunction induces, for every pair consisting of a left approximation (k, a) of f and a right approximation (k', a') of f', a **derived adjunction** $\operatorname{Ho} k \colon \operatorname{Ho} \boldsymbol{X} \leftrightarrow \operatorname{Ho} \boldsymbol{Y} \colon \operatorname{Ho} k'$ (8.4) which is natural in (k, a) and (k', a').

(iv) Given two composable deformable adjunctions $f_1 \colon \boldsymbol{X} \leftrightarrow \boldsymbol{Y} \colon f_1'$ and $f_2 \colon \boldsymbol{Y} \leftrightarrow \boldsymbol{Z} \colon f_2'$, a *sufficient* condition in order that their composition $f_2 f_1 \colon \boldsymbol{X} \leftrightarrow \boldsymbol{Z} \colon f_1' f_2'$ is also a deformable adjunction and that the compositions of their derived adjunctions are derived adjunctions of their composition is that

(iv)' the pair (f_1, f_2) is *left deformable* (ii) and the pair (f_2', f_1') is *right deformable*,

which in particular is the case if

(iv)'' the category $\boldsymbol{X}$ is *saturated* (8.4), the pair (f_1, f_2) is *left deformable* and the pair (f_2', f_1') is *locally right deformable* (which is a weaker requirement than being right deformable)

or dually if

(iv)''' the category $\boldsymbol{Z}$ is *saturated*, the pair (f_2', f_1') is *right deformable* and the pair (f_1, f_2) is just *locally left deformable*.

(v) We end with noting that, for a deformable adjunction $f\colon \boldsymbol{X} \leftrightarrow \boldsymbol{Y} \colon f'$, one can formulate a **Quillen condition** which implies the six statements of 12.5 and which, if the categories $\boldsymbol{X}$ and $\boldsymbol{Y}$ are *saturated* (8.4), is in turn implied by each of them.

12.7. Organization of the chapter. We start (in §13) with a brief review of the notion of *homotopical uniqueness*. Next we introduce *Quillen functors* (in §14), their *approximations* (in §15), their *derived adjunctions* (in §16) and *Quillen equivalences* (in §17), and in the last section (§18) we elaborate on the homotopical comments of 12.6.

13. Homotopical uniqueness

In this section we recall from chapter VI (§37 and §38) below some of the definitions and results on *homotopical uniqueness*, which are the homotopical analog of the categorical notion of *uniqueness up to a unique isomorphism* which we will call

13.1. Categorical uniqueness. This is the kind of uniqueness produced by universal properties and can be described as follows.

Call a category $\boldsymbol{G}$ **categorically contractible** if

  (i) the unique functor $\boldsymbol{G} \to [0]$ (where $[0]$ denotes the category consisting of a single object 0 and its identity map) is an equivalence of categories, or equivalently
  (ii) $\boldsymbol{G}$ is a non-empty groupoid in which there is exactly one isomorphism between any two of its objects.

Given a category $\boldsymbol{Y}$ and a non-empty set I, we then say that objects $Y_i \in \boldsymbol{Y}$ ($i \in I$) are **canonically isomorphic** or **categorically unique** if

  (iii) the *full* subcategory of $\boldsymbol{Y}$ spanned by these objects

or equivalently

  (iv) the **categorically full** subcategory of $\boldsymbol{Y}$ spanned by these objects, i.e. the full subcategory of $\boldsymbol{Y}$ spanned by these objects and all isomorphic ones

is *categorically contractible*.
Moreover as

  (v) *for every category, its full subcategory spanned by its initial or terminal objects is either empty or categorically contractible*

one can often verify the categorical uniqueness of a non-empty set of objects in a category $\boldsymbol{Y}$ by noting that these objects are *initial* or *terminal* objects of $\boldsymbol{Y}$. This is for instance what happens when one makes the somewhat imprecise statement that

  (vi) some objects in a category $\boldsymbol{X}$ are unique up to a unique isomorphism because they have a certain *initial* (or *terminal*) **universal property**,

as what one really means is that

  (vii) these objects together with some, often not explicitly mentioned, additional structure are *initial* (or *terminal*) objects in the category of all objects of $\boldsymbol{X}$ with such additional structure.

However not every categorical uniqueness results in this manner. For instance, the coproducts of an object A with a product of objects B and C are neither initial nor terminal objects of the category of diagrams of the form

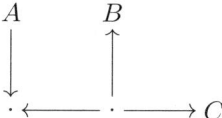

Similarly we now define

13.2. Homotopical uniqueness. We call a homotopical category $\boldsymbol{G}$ (8.2) **homotopically contractible** if

   (i) the unique functor $\boldsymbol{G} \to [0]$ (13.1(i)) is a homotopical equivalence of homotopical categories (8.3),

which in view of the two out of three property (8.2) readily implies that

   (ii) *every map in $\boldsymbol{G}$ is a weak equivalence and every two objects of $\boldsymbol{G}$ are weakly equivalent* (8.2).

Given a homotopical category $\boldsymbol{Y}$ and a non-empty set I, we then say that objects Y_i ($i \in I$) are **canonically weakly equivalent** or **homotopically unique** if

   (iii) the **homotopically full** subcategory of $\boldsymbol{Y}$ spanned by these objects, i.e. the full subcategory of $\boldsymbol{Y}$ spanned by these objects and all weakly equivalent (8.2) ones is *homotopically contractible.*

However in order to formulate the corresponding *homotopically universal properties* we first have to define appropriate

13.3. Homotopically initial and terminal objects. Given a homotopical category $\boldsymbol{Y}$, an object $Y \in \boldsymbol{Y}$ will (motivated by 38.1) be called **homotopically initial** (resp. **homotopically terminal**) if there exists a zigzag of natural transformations

$$\mathrm{cst}_Y \cdots F_0 \xrightarrow{f} F_1 \cdots 1_{\boldsymbol{Y}} \qquad (\text{resp. } 1_{\boldsymbol{Y}} \cdots F_1 \xrightarrow{f} F_0 \cdots \mathrm{cst}_Y)$$

in which

   (i) $\mathrm{cst}_Y \colon \boldsymbol{Y} \to \boldsymbol{Y}$ denotes the constant functor which sends every map of $\boldsymbol{Y}$ to the identity map of the object Y,
   (ii) $1_{\boldsymbol{Y}} \colon \boldsymbol{Y} \to \boldsymbol{Y}$ denotes the identity functor of $\boldsymbol{Y}$,
   (iii) the ·'s denote (possibly empty) zigzags of *natural weak equivalences*, and
   (iv) the map $fY \in \boldsymbol{Y}$ is a *weak equivalence*

which readily implies that, as one would expect

   (v) *for every homotopically initial or terminal object of $\boldsymbol{Y}$ its image in $\mathrm{Ho}\,\boldsymbol{Y}$ (8.4) is initial or terminal.*

Furthermore, using the two out of six property (8.2) one then (see 38.3 and 38.5) can prove that

   (vi) *if Y is a homotopically initial or terminal object of $\boldsymbol{Y}$, then an object $Y' \in \boldsymbol{Y}$ is also homotopically initial or terminal iff Y' is weakly equivalent (8.2) to Y,*

and from this it is not difficult to deduce that

(vii) *for every homotopical category, its full subcategory spanned by its homotopically initial or terminal objects is either empty or homotopically contractible*

This in turn implies that, as in the categorical case (13.1), one can often verify the homotopical uniqueness (13.2) of a non-empty set of objects in a category $\boldsymbol{Y}$ by noting that these objects are homotopically initial or homotopically terminal objects of $\boldsymbol{Y}$. It is therefore sometimes convenient to make the somewhat imprecise statement that

(viii) some objects in a homotopical category $\boldsymbol{X}$ are homotopically unique because they have a certain *homotopically initial* (or *terminal*) **homotopically universal property**,

as a shorthand for saying that

(ix) these objects together with some, not necessarily explicitly mentioned, additional structure are *homotopically initial* (or *terminal*) objects in the category of all objects of $\boldsymbol{X}$ with such additional structure.

We end with noting that an easily verifiable and sometimes useful application of this is (38.4)

13.4. Proposition. *Given a homotopical category $\boldsymbol{X}$, subcategories $\boldsymbol{X}_1, \boldsymbol{X}_2 \subset \boldsymbol{X}$ of which $\boldsymbol{X}_1$ (resp. $\boldsymbol{X}_2$) is categorically contractible (13.1), and objects $X_1 \in \boldsymbol{X}_1$ and $X_2 \in \boldsymbol{X}_2$,*

(i) *a map $X_1 \to X_2 \in \boldsymbol{X}$ is a homotopically initial (resp. terminal) object of the category $(\boldsymbol{X}_1 \downarrow \boldsymbol{X}_2)$ (which has as objects the maps $Y_1 \to Y_2 \in \boldsymbol{X}$ with $Y_1 \in \boldsymbol{X}_1$ and $Y_2 \in \boldsymbol{X}_2$), iff*

(ii) *it is a homotopically initial (resp. terminal) object of its subcategory $(X_1 \downarrow \boldsymbol{X}_2)$ (resp. $(\boldsymbol{X}_1 \downarrow X_2)$).*

14. Quillen functors

We now introduce a useful notion of "morphism between model categories" which is *not*, as one might expect, a functor which is compatible with the model structures in the sense that it preserves cofibrations, fibrations and weak equivalences, but an *adjoint pair* of functors of which the *left adjoint* is compatible with one half of the model structures in the sense that it preserves *cofibrations and trivial cofibrations*, while the *right adjoint* is compatible with the other half and preserves *fibrations and trivial fibrations*. Neither adjoint is thus required to preserve weak equivalences.

14.1. Quillen functors. Given two model categories $\boldsymbol{M}$ and $\boldsymbol{N}$ (9.1), a **Quillen adjunction** will be an adjunction

$$f\colon \boldsymbol{M} \longleftrightarrow \boldsymbol{N} : f'$$

of which

(i) the left adjoint is a **left Quillen functor**, i.e. a functor which preserves *cofibrations* and *trivial cofibrations*, and

(ii) the right adjoint is a **right Quillen functor**, i.e. a functor which preserves *fibrations* and *trivial fibrations*.

They have the

14.2. Elementary properties.

(i) *The identity functor of a model category is both a left and a right Quillen functor.*
(ii) *Every right adjoint of a left Quillen functor is a right Quillen functor and every left adjoint of a right Quillen functor is a left Quillen functor.*
(iii) *The opposite of a left Quillen functor is a right Quillen functor and the opposite of a right Quillen functor is a left Quillen functor.*
(iv) *The composition of two left Quillen functors is a left Quillen functor and the composition of two right Quillen functors is a right Quillen functor.*

Proof. Parts (i) and (iv) are obvious, part (iii) follows from 9.7(i) and part (ii) can be verified using 9.5.

14.3. Remark. Probably the most useful of the above properties is 14.2(ii) as, for an adjoint pair of functors $f\colon \boldsymbol{M} \leftrightarrow \boldsymbol{N} : f'$ between model categories, of the two equivalent statements

(i) *f preserves cofibrations and trivial cofibrations*, and
(ii) *f' preserves fibrations and trivial fibrations*

one is often much easier to verify than the other.

Less obvious is the following

14.4. Deformability result for Quillen functors. *Let $f\colon \boldsymbol{M} \leftrightarrow \boldsymbol{N} : f'$ be a Quillen adjunction (14.1). Then (8.3, 10.2 and 10.3)*

(i) *f is homotopical on the left deformation retract $\boldsymbol{M}_c$ of $\boldsymbol{M}$, i.e. f preserves weak equivalences between cofibrant objects, and dually*
(ii) *f' is homotopical on the right deformation retract $\boldsymbol{N}_f$ of $\boldsymbol{N}$, i.e. f' preserves weak equivalences between fibrant objects.*

Proof. This follows readily from

14.5. Ken Brown's lemma. *Let $\boldsymbol{C}$ be a homotopical category and let $\boldsymbol{M}$ be a model category. Then every functor $\boldsymbol{M} \to \boldsymbol{C}$ which sends trivial cofibrations (resp. trivial fibrations) to weak equivalences in $\boldsymbol{C}$, sends weak equivalences in $\boldsymbol{M}_c$ (resp. $\boldsymbol{M}_f$) (10.2) to weak equivalences in $\boldsymbol{C}$, i.e. is homotopical (8.3) on $\boldsymbol{M}_c$ (resp. $\boldsymbol{M}_f$).*

Proof (of the cofibration half). Given a weak equivalence $f\colon A \to B \in \boldsymbol{M}_c$, factor the map $(f, 1_B)\colon A \amalg B \to B \in \boldsymbol{M}_c$ into a cofibration $j\colon A \amalg B \rightarrowtail C$ and a trivial fibration $p\colon C \xrightarrow{\sim}\twoheadrightarrow B$. Then the composite cofibrations

$$A \xrightarrowtail{i_1} A \amalg B \xrightarrowtail{j} C \qquad \text{and} \qquad B \xrightarrowtail{i_2} A \amalg B \xrightarrowtail{j} C$$

have the property that $p(ji_1) = f$ and $p(ji_2) = 1_B$ and therefore are, in view of the two out of three property, trivial cofibrations and the desired result now follows readily using again the two out of three property.

15. Approximations

The usefulness of left and right Quillen functors is due to the fact that they have *left* and *right approximations*, i.e. homotopically unique (13.2) *homotopical* functors which, in a homotopical sense, are closest to them from the left or from the right.

To define these approximations it is convenient to consider certain

15.1. Auxiliary categories. With a (not necessarily homotopical) functor $f\colon X \to Y$ between homotopical categories (8.2 and 8.3) one can associate the **homotopical category of the homotopical functors $X \to Y$ over** (resp. **under**) f, i.e. the homotopical category

$$\big(\mathrm{Fun}_w(X, Y) \downarrow f\big) \qquad (\text{resp. } \big(f \downarrow \mathrm{Fun}_w(X, Y)\big))$$

which has

(i) as objects the pairs (k, a), where k is a homotopical functor $k\colon X \to Y$ and a is a natural transformation

$$a\colon k \longrightarrow f \qquad (\text{resp. } a\colon f \longrightarrow k)$$

and which has

(ii) as maps and weak equivalences $t\colon (k_1, a_1) \to (k_2, a_2)$ respectively the natural transformations and natural weak equivalences $t\colon k_1 \to k_2$ such that

$$a_2 t = a_1 \qquad (\text{resp. } t a_1 = a_2).$$

15.2. Approximations of Quillen functors. Given a Quillen adjunction $f\colon M \leftrightarrow N :f'$

(i) a **left approximation** of the left Quillen functor f will be a *homotopically terminal* object (13.3) of $\big(\mathrm{Fun}_w(M, N) \downarrow f\big)$ (15.1), and
(ii) a **right approximation** of the right Quillen functor f' will be a *homotopically initial* object of $\big(f' \downarrow \mathrm{Fun}_w(N, M)\big)$.

Thus (13.3(viii))

(iii) *the left approximations of f and the right approximations of f', if they exist, are homotopically unique* (13.2)

and it remains to verify the

15.3. Existence of approximations of Quillen functors. *For every Quillen adjunction $f\colon M \leftrightarrow N :f'$ (14.1), left deformation (r, s) of M into M_c (8.3 and 10.2) and right deformation (r', s') of N into N_f, the pairs*

$$(fr, fs) \qquad \text{and} \qquad (f'r', f's')$$

are respectively a left approximation of f (15.2) and a right approximation of f'.

Proof (of the left half). In view of 14.4 the pair (fr, fs) is an object of the category $\big(\mathrm{Fun}_w(M, N) \downarrow f\big)$ and to show that this object is homotopically terminal (13.3) one notes that every object $(k, a) \in \big(\mathrm{Fun}_w(M, N) \downarrow f\big)$ gives rise to a zigzag in $\big(\mathrm{Fun}_w(M, N) \downarrow f\big)$

$$(k, a) \xleftarrow{ks} (kr, as) \xrightarrow{ar} (fr, fs)$$

where as denotes the diagonal of the commutative square

$$\begin{array}{ccc} kr & \xrightarrow{ar} & fr \\ ks \downarrow & & \downarrow fs \\ k & \xrightarrow{a} & f \end{array}$$

which has the property that
  (i) this zigzag is natural in (k, a),
  (ii) the map ks is a weak equivalence, and
  (iii) if $(k, a) = (fr, fs)$, then $ar = fsr$ is also a weak equivalence.

As one would like, approximations are compatible with compositions, and to prove this we first consider two

15.4. Composition functors. Given two composable (not necessarily homotopical) functors

$$f_1 \colon X \longrightarrow Y \quad \text{and} \quad f_2 \colon Y \longrightarrow Z$$

between homotopical categories (8.2 and 8.3), one can consider the **composition functor** (15.1)

$$\bigl(\mathrm{Fun}_w(X, Y) \downarrow f_1\bigr) \times \bigl(\mathrm{Fun}_w(Y, Z) \downarrow f_2\bigr) \longrightarrow \bigl(\mathrm{Fun}_w(X, Z) \downarrow f_2 f_1\bigr)$$

which sends a pair of objects (k_1, a_1) and (k_2, a_2) to their **composition**

$$(k_2, a_2)(k_1, a_1) = (k_2 k_1, a_2 a_1)$$

where $a_2 a_1$ denotes the diagonal of the commutative diagram

$$\begin{array}{ccc} k_2 k_1 & \xrightarrow{a_2} & f_2 k_1 \\ a_1 \downarrow & & \downarrow a_1 \\ k_2 f_1 & \xrightarrow{a_2} & f_2 f_1 \end{array}$$

and the **composition functor**

$$\bigl(f_1 \downarrow \mathrm{Fun}_w(X, Y)\bigr) \times \bigl(f_2 \downarrow \mathrm{Fun}_w(Y, Z)\bigr) \longrightarrow \bigl(f_2 f_1 \downarrow \mathrm{Fun}_w(X, Z)\bigr)$$

which sends a pair of objects (k_1, a_1) and (k_2, a_2) to their **composition**

$$(k_2, a_2)(k_1, a_1) = (k_2 k_1, a_2 a_1)$$

where $a_2 a_1$ denotes the diagonal of the commutative diagram

$$\begin{array}{ccc} f_2 f_1 & \xrightarrow{a_2} & k_2 f_1 \\ a_1 \downarrow & & \downarrow a_1 \\ f_2 k_1 & \xrightarrow{a_2} & k_2 k_1 \end{array}$$

and note that
  (i) *these composition functors are both homotopical* (8.3)
and hence (13.3(vi))
  (ii) *if they send one homotopically terminal (or initial) object to a homotopically terminal (or initial) one, then they do so for every homotopically terminal (or initial) object.*

Now we can state the

15.5. Composability of approximations of Quillen functors. *Given two composable Quillen adjunctions*

$$f_1\colon \boldsymbol{M} \longleftrightarrow \boldsymbol{N} :f_1' \quad \text{and} \quad f_2\colon \boldsymbol{N} \longleftrightarrow \boldsymbol{P} :f_2'$$

(i) *their composition*

$$f_2 f_1\colon \boldsymbol{M} \longleftrightarrow \boldsymbol{P} :f_1' f_2'$$

is also a Quillen adjunction,
(ii) *every composition of a left approximation of f_1 with a left approximation of f_2 is a left approximation of $f_2 f_1$, and*
(iii) *every composition of a right approximation of f_2' with a right approximation of f_1' is a right approximation of $f_1' f_2'$.*

Proof (of the left half). Part (i) is an immediate consequence of 14.2(iv) and in view of 15.4(ii) it suffices to verify (ii) for only one choice of left deformation. To do this, let (r_1, s_1) and (r_2, s_2) be left deformations of $\boldsymbol{M}$ into $\boldsymbol{M}_c$ and $\boldsymbol{N}$ into $\boldsymbol{N}_c$ respectively (8.3 and 10.2). Then the map (15.3 and 15.4)

$$(f_2 r_2, f_2 s_2)(f_1 r_1, f_1 s_1) \xrightarrow{f_2 s_2 f_1 r_1} (f_2 f_1 r_1, f_2 f_1 s_1) \in (\mathrm{Fun}_{\mathrm{w}}(\boldsymbol{M}, \boldsymbol{P}) \downarrow f_2 f_1)$$

is readily verified to be a weak equivalence (15.1) and hence (13.3(vi)) the composition $(f_2 r_2, f_2 s_2)(f_1 r_1, f_1 s_1)$ (15.4) is just like $(f_2 f_1 r_1, f_2 f_1 s_1)$ a left approximation of $f_2 f_1$.

16. Derived adjunctions

A useful property of the approximations of Quillen functors is that
  (i) for every Quillen adjunction $f\colon \boldsymbol{M} \leftrightarrow \boldsymbol{N} :f'$, its adjunction induces, for every pair consisting of a left approximation (k, a) of f and a right approximation (k', a') of f' (15.2), a *derived adjunction*

$$\mathrm{Ho}\, k\colon \mathrm{Ho}\,\boldsymbol{M} \longleftrightarrow \mathrm{Ho}\,\boldsymbol{N} :\mathrm{Ho}\, k'$$

  which is *natural* in (k, a) and (k', a'), and has the property that
  (ii) *for every two composable Quillen adjunctions, the compositions of their derived adjunctions are derived adjunctions of their composition.*

Before precisely formulating this result we first note the existence of

16.1. An induced partial adjunction. *Let $f\colon \boldsymbol{M} \leftrightarrow \boldsymbol{N} :f'$ be a Quillen adjunction. Then, for every pair of objects $X \in \boldsymbol{M}_c$ and $Y \in \boldsymbol{N}_f$ (10.2), the adjunction isomorphism*

$$\boldsymbol{N}(fX, Y) \approx \boldsymbol{M}(X, f'Y)$$

is compatible with the homotopy relation $\sim$ (10.5(iii)) and hence (10.6) induces a **partial adjunction isomorphism** *(10.1)*

$$\mathrm{Ho}\,\boldsymbol{N}(fX, Y) \stackrel{\phi}{\approx} \mathrm{Ho}\,\boldsymbol{M}(X, f'Y)$$

which is natural in X and Y.

16. DERIVED ADJUNCTIONS

Moreover, for every two composable Quillen adjunctions $f_1 \colon \boldsymbol{M} \leftrightarrow \boldsymbol{N} \colon f_1'$ and $f_2 \colon \boldsymbol{N} \leftrightarrow \boldsymbol{P} \colon f_2'$ and pair of objects $X \in \boldsymbol{M}_c$ and $Z \in \boldsymbol{P}_f$, the diagram of partial adjunction isomorphisms

$$\begin{array}{ccc}
\operatorname{Ho}\boldsymbol{P}(f_2 f_1 X, Z) & \xrightarrow[\approx]{\phi} & \operatorname{Ho}\boldsymbol{M}(X, f_1' f_2' Z) \\
& \searrow_{\approx}^{\phi} \quad \nearrow_{\approx}^{\phi} & \\
& \operatorname{Ho}\boldsymbol{N}(f_1 X, f_2' Z) &
\end{array}$$

commutes.

Proof. This follows from the observation that, as (in view of 14.4) the functor f' sends the factorization (10.5(ii)) $Y \xrightarrow{\sim} Y' \twoheadrightarrow Y \times Y$ of the diagonal map $Y \to Y \times Y$ to a similar factorization of the diagonal map $f'Y \to f'(Y \times Y) = f'Y \times f'Y$, the given adjunction sends right homotopic maps $fX \to Y \in \boldsymbol{N}$ to right homotopic maps $X \to f'Y \in \boldsymbol{M}$ and left homotopic maps $X \to f'Y \in \boldsymbol{M}$ to left homotopic maps $fX \to Y \in \boldsymbol{N}$, and the second part then follows readily from the first.

We also need certain

16.2. Canonical natural isomorphisms. *Let $f \colon \boldsymbol{M} \leftrightarrow \boldsymbol{N} \colon f'$ be a Quillen adjunction. Then there are unique functions u and u' which assign to every pair of left approximations (k_1, a_1) and (k_2, a_2) of f and every pair of right approximations (k_1', a_1') and (k_2', a_2') of f', natural isomorphisms (called **canonical natural isomorphisms**)*

$$u(k_1, k_2) \colon \operatorname{Ho} k_1 \longrightarrow \operatorname{Ho} k_2 \quad \text{and} \quad u'(k_2', a_2') \colon \operatorname{Ho} k_2' \longrightarrow \operatorname{Ho} k_1'$$

such that

(i) *u and u' are natural in both variables, and*
(ii) *for every map*

$$t \colon (k_1, a_1) \longrightarrow (k_2, a_2) \in \bigl(\operatorname{Fun}_{\mathrm{w}}(\boldsymbol{M}, \boldsymbol{N}) \downarrow f\bigr)$$

and every map

$$t' \colon (k_2', a_2') \longrightarrow (k_1', a_1') \in \bigl(f' \downarrow \operatorname{Fun}_{\mathrm{w}}(\boldsymbol{N}, \boldsymbol{M})\bigr)$$

one has

$$u(k_1, k_2) = \operatorname{Ho} t \quad \text{and} \quad u'(k_2', k_1') = \operatorname{Ho} t' \ .$$

Proof (of the left half). The composition

$$\bigl(\operatorname{Fun}_{\mathrm{w}}(\boldsymbol{M}, \boldsymbol{N}) \downarrow f\bigr) \xrightarrow{j} \operatorname{Fun}_{\mathrm{w}}(\boldsymbol{M}, \boldsymbol{N}) \xrightarrow{\operatorname{Ho}} \operatorname{Fun}(\operatorname{Ho}\boldsymbol{M}, \operatorname{Ho}\boldsymbol{N})$$

in which j denotes the forgetful map and Ho is as in 8.4, sends weak equivalences to isomorphisms and hence (8.4) admits a unique factorization

$$\bigl(\operatorname{Fun}_{\mathrm{w}}(\boldsymbol{M}, \boldsymbol{N}) \downarrow f\bigr) \xrightarrow{\gamma'} \operatorname{Ho}\bigl(\operatorname{Fun}_{\mathrm{w}}(\boldsymbol{M}, \boldsymbol{N}) \downarrow f\bigr) \longrightarrow \operatorname{Fun}(\operatorname{Ho}\boldsymbol{M}, \operatorname{Ho}\boldsymbol{N})$$

in which γ' denotes the localization functor. One then readily verifies that the function which sends a pair $\bigl((k_1, a_2), (k_2, a_2)\bigr)$ to the image in $\operatorname{Fun}(\operatorname{Ho}\boldsymbol{M}, \operatorname{Ho}\boldsymbol{N})$ of the (in view of 13.3(v)) *unique* map

$$\gamma'(k_1, a_1) \longrightarrow \gamma'(k_2, a_2) \in \operatorname{Ho}\bigl(\operatorname{Fun}_{\mathrm{w}}(\boldsymbol{M}, \boldsymbol{N}) \downarrow f\bigr)$$

has the desired properties.

Using these canonical natural isomorphisms we now implicitly define

16.3. Derived adjunctions of Quillen adjunctions. *Given a Quillen adjunction* $f\colon \boldsymbol{M} \leftrightarrow \boldsymbol{N} :f'$, *there is a unique function which associates with every pair consisting of a left approximation* (k,a) *of* f (15.2) *and a right approximation* (k',a') *of* f', *a **derived adjunction***

$$\operatorname{Ho} k\colon \operatorname{Ho} \boldsymbol{M} \longleftrightarrow \operatorname{Ho} \boldsymbol{N} :\operatorname{Ho} k'$$

such that

(i) *for every two left approximations* (k_1,a_1) *and* (k_2,a_2) *of* f *and right approximations* (k'_1,a'_1) *and* (k'_2,a'_2) *of* f', *the canonical natural isomorphisms* (16.2)

$$u(k_1,k_2)\colon \operatorname{Ho} k_1 \longrightarrow \operatorname{Ho} k_2 \qquad \text{and} \qquad u'(k'_2,k'_1)\colon \operatorname{Ho} k'_2 \longrightarrow \operatorname{Ho} k'_1$$

are conjugate (39.10) *with respect to the associated derived adjunctions*

$$\operatorname{Ho} k_1\colon \operatorname{Ho} \boldsymbol{M} \longleftrightarrow \operatorname{Ho} \boldsymbol{N} :\operatorname{Ho} k'_1 \qquad \text{and} \qquad \operatorname{Ho} k_2\colon \operatorname{Ho} \boldsymbol{M} \longleftrightarrow \operatorname{Ho} \boldsymbol{N} :\operatorname{Ho} k'_2$$

and

(ii) *for every left deformation* (r,s) *of* $\boldsymbol{M}$ *into* $\boldsymbol{M}_c$ (8.3) *and right deformation* (r',s') *of* $\boldsymbol{N}$ *into* $\boldsymbol{N}_f$, *the adjunction isomorphism of the derived adjunction*

$$\operatorname{Ho}(fr)\colon \operatorname{Ho} \boldsymbol{M} \longleftrightarrow \operatorname{Ho} \boldsymbol{N} :\operatorname{Ho}(f'r')$$

associated with (fr, fs) *and* $(f'r', f's')$ (15.3) *assigns to every pair of objects* $X \in \boldsymbol{M}$ *and* $Y \in \boldsymbol{N}$ *the composition*

$$\operatorname{Ho} \boldsymbol{N}(frX, Y) \approx \operatorname{Ho} \boldsymbol{N}(frX, r'Y) \overset{\phi}{\approx} \operatorname{Ho} \boldsymbol{M}(rX, f'r'Y) \approx \operatorname{Ho} \boldsymbol{M}(X, f'r'Y)$$

in which ϕ *is the partial adjunction isomorphism* (16.1) *and the other isomorphisms are induced by* s' *and* s.

16.4. Corollary. *Let* $f\colon \boldsymbol{M} \leftrightarrow \boldsymbol{N} :f'$ *be a Quillen adjunction and let* (k,a) *and* (k',a') *respectively be a left approximation of* f *and a right approximation of* f'. *Then the functor* $\operatorname{Ho} k\colon \operatorname{Ho} \boldsymbol{M} \to \operatorname{Ho} \boldsymbol{N}$ *is an equivalence of categories iff the functor* $\operatorname{Ho} k'\colon \operatorname{Ho} \boldsymbol{N} \to \operatorname{Ho} \boldsymbol{M}$ *is so.*

We also note that the above (16.3) derived adjunctions are compatible with

16.5. Composability of derived adjunctions of Quillen adjunctions. *Let*

$$f_1\colon \boldsymbol{M} \longleftrightarrow \boldsymbol{N} :f'_1 \qquad \text{and} \qquad f_2\colon \boldsymbol{N} \longleftrightarrow \boldsymbol{P} :f'_2$$

be two composable Quillen adjunctions and let

$$(k_1,a_1) \text{ and } (k_2,a_2) \qquad \text{and} \qquad (k'_1,a'_1) \text{ and } (k'_2,a'_2)$$

respectively be left approximations of f_1 *and* f_2 *and right approximations of* f'_1 *and* f'_2. *Then the composition of the associated derived adjunctions*

$$\operatorname{Ho} k_1\colon \operatorname{Ho} \boldsymbol{M} \longleftrightarrow \operatorname{Ho} \boldsymbol{N} :\operatorname{Ho} k'_1 \qquad \text{and} \qquad \operatorname{Ho} k_2\colon \operatorname{Ho} \boldsymbol{N} \longleftrightarrow \operatorname{Ho} \boldsymbol{P} :\operatorname{Ho} k'_2$$

is exactly the derived adjunction

$$\operatorname{Ho}(k_2 k_1)\colon \operatorname{Ho} \boldsymbol{M} \longleftrightarrow \operatorname{Ho} \boldsymbol{P} :\operatorname{Ho}(k'_1 k'_2)$$

associated with their compositions (15.4 and 15.5)

$$(k_2k_1, a_2a_1) \quad \text{and} \quad (k'_1k'_2, a'_1a'_2)$$

It thus remains to give a proof of 16.3 and 16.5.

16.6. Proof of 16.3. As (39.10)

(i) given an adjunction $g_1 \colon \boldsymbol{X} \leftrightarrow \boldsymbol{Y} : g'_1$, a pair of functors $g_2 \colon \boldsymbol{X} \to \boldsymbol{Y}$ and $g'_2 \colon \boldsymbol{Y} \to \boldsymbol{X}$ and a pair of natural *isomorphisms* $h \colon g_1 \to g_2$ and $h' \colon g'_2 \to g'_1$, there is a *unique* adjunction $g_2 \colon \boldsymbol{X} \leftrightarrow \boldsymbol{Y} : g'_2$ such that h and h' are conjugate with respect to these two adjunctions

it suffices, in view of 16.2, to show that

(ii) for every two left deformations (r_1, s_1) and (r_2, s_2) of $\boldsymbol{M}$ into $\boldsymbol{M}_c$ and right deformations (r'_1, s'_1) and (r'_2, s'_2) of $\boldsymbol{N}$ into $\boldsymbol{N}_f$, the canonical natural isomorphisms (16.2)

$$u(fr_1, fr_2) \colon \operatorname{Ho}(fr_1) \longrightarrow \operatorname{Ho}(fr_2)$$

$$\text{and} \quad u'(f'r'_2, f'r'_1) \colon \operatorname{Ho}(f'r'_2) \longrightarrow \operatorname{Ho}(f'r'_1)$$

associated with the approximations (15.3)

$$(fr_1, fs_1) \text{ and } (fr_2, fs_2) \quad \text{and} \quad (f'r'_2, f's'_2) \text{ and } (f'r'_1, f's'_1)$$

are conjugate with respect to the derived adjunctions

$$\operatorname{Ho}(fr_1) \colon \operatorname{Ho}\boldsymbol{M} \longleftrightarrow \operatorname{Ho}\boldsymbol{N} : \operatorname{Ho}(f'r'_1)$$

$$\text{and} \quad \operatorname{Ho}(fr_2) \colon \operatorname{Ho}\boldsymbol{M} \longleftrightarrow \operatorname{Ho}\boldsymbol{N} : \operatorname{Ho}(f'r'_2)$$

associated, as in 16.3(ii), with the approximations

$$(fr_1, fs_1) \text{ and } (f'r'_1, f's'_1) \quad \text{and} \quad (fr_2, fs_2) \text{ and } (f'r'_2, f's'_2) \ .$$

To do this one notes that the commutativity of the diagrams

$$\begin{array}{ccc} r_2r_1 & \xrightarrow{r_2s_1} & r_2 \\ {\scriptstyle s_2r_1}\downarrow & & \downarrow{\scriptstyle s_2} \\ r_1 & \xrightarrow{s_1} & 1_M \end{array} \quad \text{and} \quad \begin{array}{ccc} 1_N & \xrightarrow{s'_1} & r'_1 \\ {\scriptstyle s'_2}\downarrow & & \downarrow{\scriptstyle s'_2r'_1} \\ r'_2 & \xrightarrow{r'_2s'_1} & r'_2r'_1 \end{array}$$

implies that

$$(r_2r_1, s_2s_1) \quad \text{and} \quad (r'_2r'_1, s'_2s'_1)$$

where s_2s_1 and $s'_2s'_1$ denote their diagonals, are a left deformation of $\boldsymbol{M}$ into $\boldsymbol{M}_c$ and a right deformation of $\boldsymbol{N}$ into $\boldsymbol{N}_f$, from which in turn one can deduce that (15.3 and 16.2)

$$\operatorname{Ho}(fr_2s_1)\operatorname{Ho}(fs_2r_1)^{-1} = u(fr_1, fr_2) \colon \operatorname{Ho}(fr_1) \longrightarrow \operatorname{Ho}(fr_2)$$

and

$$\operatorname{Ho}(f's'_2r'_1)^{-1}\operatorname{Ho}(f'r'_2s'_1) = u'(f'r'_2, f'r'_1) \colon \operatorname{Ho}(f'r'_2) \longrightarrow \operatorname{Ho}(f'r'_1)$$

and the desired result then follows from the observation that every pair of objects $X \in \boldsymbol{M}$ and $Y \in \boldsymbol{N}$ gives rise to a commutative diagram of isomorphisms

$$\begin{array}{ccc}
\operatorname{Ho} \boldsymbol{N}(fr_2 X, Y) \approx \cdots \overset{\phi}{\approx} \cdots \approx \operatorname{Ho} \boldsymbol{M}(X, f'r_2' Y) \\
{\scriptstyle (fr_2 s_1)^*}\downarrow \qquad\qquad\qquad\qquad\qquad \downarrow{\scriptstyle (f' r_2' s_1')_*} \\
\operatorname{Ho} \boldsymbol{N}(fr_2 r_1 X, Y) \approx \cdots \overset{\phi}{\approx} \cdots \approx \operatorname{Ho} \boldsymbol{M}(X, f' r_2' r_1' Y) \\
{\scriptstyle (fs_2 r_1)^*}\uparrow \qquad\qquad\qquad\qquad\qquad \uparrow{\scriptstyle (f' s_2' r_1')_*} \\
\operatorname{Ho} \boldsymbol{N}(fr_1 X, Y) \approx \cdots \overset{\phi}{\approx} \cdots \approx \operatorname{Ho} \boldsymbol{M}(X, f' r_1' Y)
\end{array}$$

in which the horizontal sequences are as in 16.3(ii).

16.7. Proof of 16.5. In view of 16.3(i) it suffices to prove 16.5 for only one choice of the left and right approximations. To do this let

$$(r_1, s_1) \text{ and } (r_2, s_2) \quad \text{and} \quad (r_1', s_1') \text{ and } (r_2', s_2')$$

respectively be left approximations of $\boldsymbol{M}$ into $\boldsymbol{M}_c$ and $\boldsymbol{N}$ into $\boldsymbol{N}_c$ and right approximations of $\boldsymbol{N}$ into $\boldsymbol{N}_f$ and $\boldsymbol{P}$ into $\boldsymbol{P}_f$. Then the desired result follows readily from the observation that, for every pair of objects $X \in \boldsymbol{M}$ and $Z \in \boldsymbol{P}$, in view of 16.1 and the naturality of the maps induced by s_1, s_2, s_1' and s_2', the following diagram of isomorphisms commutes.

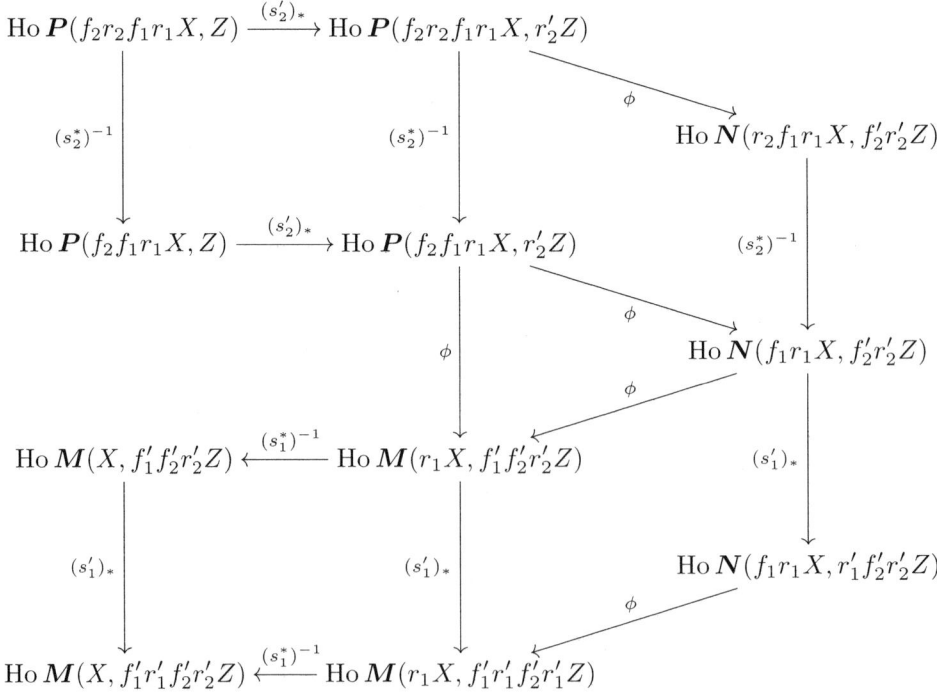

17. Quillen equivalences

We now briefly discuss *Quillen equivalences*, i.e. Quillen adjunctions $f\colon \boldsymbol{M} \leftrightarrow \boldsymbol{N} :f'$ (14.1) with the property that, for every left approximation (k, a) of f (15.2) and every right approximation (k', a') of f', the functors

$$k\colon \boldsymbol{M} \longrightarrow \boldsymbol{N} \quad \text{and} \quad k'\colon \boldsymbol{N} \longrightarrow \boldsymbol{M}$$

are *inverse homotopical equivalences of homotopical categories* (8.3).

We start with noting that corollary 16.4 generalizes to

17.1. Proposition. *Let* $f\colon \boldsymbol{M} \leftrightarrow \boldsymbol{N} \colon f'$ *be a Quillen adjunction* (14.1). *Then the following six statements are equivalent:*

(i) *for one (and hence every) left approximation* (15.2) (k, a) *of f and one (and hence every) right approximation* (k', a') *of f', the homotopical functors k and k' are homotopically inverse homotopical equivalences of categories* (8.3)

(i)′ *for one (and hence every) left approximation* (k, a) *of f, the homotopical functor k is a homotopical equivalence of homotopical categories,*

(i)″ *for one (and hence every) right approximation* (k', a') *of f' the homotopical functor k' is a homotopical equivalence of homotopical categories,*

(ii) *for one (and hence every) left approximation* (k, a) *of f and one (and hence every) right approximation* (k', a') *of f', the functors $\mathrm{Ho}\, k$ and $\mathrm{Ho}\, k'$ are inverse equivalences of categories*

(ii)′ *for one (and hence every) left approximation* (k, a) *of f, the functor $\mathrm{Ho}\, k$ is an equivalence of categories, and*

(ii)″ *for one (and hence every) right approximation* (k', a') *of f', the functor $\mathrm{Ho}\, k'$ is an equivalence of categories.*

A Quillen adjunction with these properties thus induces some kind of "equivalence of homotopy theories", and such a pair is therefore called an adjunction of

17.2. Quillen equivalences. A Quillen adjunction with the six equivalent properties of (17.2) will be called an **adjunction of Quillen equivalences** and one refers to its left Quillen functor as a **left Quillen equivalence** and its right Quillen functor as a **right Quillen equivalence**.

A useful *necessary and sufficient* condition in order that a Quillen adjunction be an adjunction of Quillen equivalences is the so-called

17.3. Quillen condition. *A Quillen adjunction* $f\colon \boldsymbol{M} \leftrightarrow \boldsymbol{N} \colon f'$ *is an adjunction of Quillen equivalences* (17.2) *iff it satisfies the* **Quillen condition** *that, for every pair of objects* $X \in \boldsymbol{M}_c$ *and* $Y \in \boldsymbol{N}_f$ (10.2), *a map* $fX \to Y \in \boldsymbol{N}$ *is a weak equivalence iff its adjunct* $X \to f'Y \in \boldsymbol{M}$ *is so.*

17.4. Corollary. *If for three Quillen adjunctions*

$$f_1\colon \boldsymbol{X} \longleftrightarrow \boldsymbol{Y} \colon f'_1, \quad f_2\colon \boldsymbol{Y} \longleftrightarrow \boldsymbol{Z} \colon f'_2 \quad \text{and} \quad f_3\colon \boldsymbol{Z} \longleftrightarrow \boldsymbol{T} \colon f'_3$$

the two compositions

$$f_2 f_1\colon \boldsymbol{X} \longleftrightarrow \boldsymbol{Z} \colon f'_1 f'_2 \quad \text{and} \quad f_3 f_2\colon \boldsymbol{Y} \longleftrightarrow \boldsymbol{T} \colon f'_2 f'_3$$

are adjunctions of Quillen equivalences, then so are the original three.

It remains to give

Proofs of 17.1 *and* 17.3. In view of the fact that *a functor is an equivalence of categories iff it is part of an adjunction of which the unit and the counit are natural isomorphisms,* 17.1 and 17.3 follow readily from 16.4 and the following

17.5. Proposition. *Let* $f\colon \mathbf{M} \leftrightarrow \mathbf{N} : f'$ *be a Quillen adjunction and let* (r,s) *and* (r',s') *respectively be a left deformation of* $\mathbf{M}$ *into* $\mathbf{M}_c$ *and a right deformation of* $\mathbf{N}$ *into* $\mathbf{N}_f$ *(8.3 and 10.2). Then the following three statements are equivalent:*

(i) *for every pair of objects* $X \in \mathbf{M}_c$ *and* $Y \in \mathbf{N}_f$, *a map* $X \to f'Y \in \mathbf{M}$ *is a weak equivalence if (resp. only if) its adjunct* $fX \to Y \in \mathbf{N}$ *is so*

(ii) *the zigzag of natural transformations*

$$1_{\mathbf{M}} \xleftarrow{s} r \longrightarrow f'r'fr \qquad (\text{resp. } frf'r' \longrightarrow r' \xleftarrow{s'} 1_{\mathbf{N}})$$

in which the unnamed map is the adjunct of the natural weak equivalence

$$fr \xrightarrow{s'fr} r'fr \qquad (\text{resp. } rf'r' \xrightarrow{sf'r'} f'r')$$

is a zigzag of natural weak equivalences, and

(iii) *the unit (resp. counit) (39.10)*

$$1_{\operatorname{Ho} \mathbf{M}} \longrightarrow \operatorname{Ho}(f'r')\operatorname{Ho}(fr) \qquad (\text{resp. } \operatorname{Ho}(fr)\operatorname{Ho}(f'r') \longrightarrow 1_{\operatorname{Ho} \mathbf{N}})$$

of the associated derived adjunction (16.3(ii))

$$\operatorname{Ho}(fr)\colon \operatorname{Ho} \mathbf{M} \longleftrightarrow \operatorname{Ho} \mathbf{N} :\operatorname{Ho}(f'r')$$

is a natural isomorphism.

Proof (of the first half). Clearly (i) implies (ii) and a straightforward calculation using 16.1, 16.3(ii) and 39.10 yields that the image of the zigzag $1_{\mathbf{M}} \xleftarrow{s} r \to f'r'fr$ in $\operatorname{Ho} \mathbf{X}$ is exactly the unit $1_{\operatorname{Ho} \mathbf{M}} \to \operatorname{Ho}(f'r')\operatorname{Ho}(fr)$ of the derived adjunction, so that (ii) implies (iii).

It thus remains to show that (iii) implies (i) and we do this by successively noting that, if $p\colon X \to f'Y \in \mathbf{M}$ is a map such that its adjunct $q\colon fX \to Y \in \mathbf{N}$ is a weak equivalence, then

(i) the map $q(fsX)\colon frX \to Y \in \mathbf{N}$ is also a weak equivalence and hence its image $\gamma'(q(fsX)) \in \operatorname{Ho} \mathbf{N}$ under the localization functor $\gamma'\colon \mathbf{N} \to \operatorname{Ho} \mathbf{N}$ is an isomorphism,

(ii) in view of 16.1 and 16.3(ii), the adjunct of $\gamma'(q(fsX))$ is the image $\gamma((f's'Y)p) \in \operatorname{Ho} \mathbf{M}$ of the map $(f's'Y)p\colon X \to f'r'Y \in \mathbf{M}$ under the localization functor $\gamma\colon \mathbf{M} \to \operatorname{Ho} \mathbf{M}$,

(iii) this adjunct $\gamma((f's'Y)p)$ is the composition of the unit (which is assumed to be an isomorphism) with the image of $\gamma'(q(fsX))$ under the functor $\operatorname{Ho}(f'r')$ and hence is also an isomorphism, and

(iv) in view of the saturation of $\mathbf{M}$ (10.8), the map $(f's'Y)'$ is thus a weak equivalence and so is therefore the original map $p\colon X \to fY \in \mathbf{M}$.

18. Homotopical comments

In this last section we note that the main results of this chapter are special cases of more general results on homotopical categories which we will obtain in chapter VII, some of which we will need in chapter IV.

We start with observing that, in view of 14.4, Quillen functors (14.1) are

18.1. Deformable functors. Given two homotopical categories X and Y (8.2), a (not necessarily homotopical) functor $f\colon X \to Y$ is called **left** (resp. **right**) **deformable** if (40.2) f is *homotopical* (8.3) on a *left* (resp. a *right*) *deformation retract* of X (8.3).

Similarly Quillen adjunctions (14.1) are

18.2. Deformable adjunctions. Given two homotopical categories X and Y, an adjoint pair of functors $f\colon X \leftrightarrow Y : f'$ is called a **deformable adjunction** if (43.1) the *left adjoint* f is *left deformable* (18.1) and the *right adjoint* is *right deformable*.

Just like Quillen functors (15.2 and 15.3), deformable functors have

18.3. Approximations. Given a (not necessarily homotopical) functor $f\colon X \to Y$ between homotopical categories, a **left** (resp. a **right**) **approximation** of f is (41.1) a *homotopically terminal* (resp. *initial*) *object* (13.3) of the "homotopical category of homotopical functors $X \to Y$ over (resp. under) f"

$$\left(\mathrm{Fun}_w(X, Y) \downarrow f\right) \qquad (\text{resp. } \left(f \downarrow \mathrm{Fun}_w(X, Y)\right))$$

(see 15.1), and thus (13.3(viii))

(i) *such approximations of f, if they exist, are homotopically unique* (13.2).

The existence of approximations of Quillen functors (15.2 and 15.3) then is a special case of 41.2, which is proven exactly as 15.3, and which states that

(ii) a *sufficient* condition in order that a (not necessarily homotopical) functor $f\colon X \to Y$ between homotopical categories has left (resp. right) approximations is that f be *left* (resp. *right*) *deformable* (18.1),

as in that case

(iii) *for every left (resp. right) deformation retract $X_0 \subset X$ (8.3) on which f is homotopical and every left (resp. right) deformation (r, s) of X into X_0, the pair*

$$(fr, fs) \in \left(\mathrm{Fun}_w(X, Y) \downarrow f\right) \qquad (\text{resp. } (fr, fs) \in \left(f \downarrow \mathrm{Fun}_w(X, Y)\right))$$

is a left (resp. a right) approximation of f.

Next we consider

18.4. Compositions of deformable functors. Deformable functors are not as well behaved under composition as Quillen functors. For instance, the composition of two deformable functors need not be deformable, nor if it is, are the compositions of their approximations necessarily approximations of their composition. To deal with this we call, for two composable (not necessarily homotopical) functors between homotopical categories

$$f_1\colon \boldsymbol{X} \longrightarrow \boldsymbol{Y} \quad \text{and} \quad f_2\colon \boldsymbol{Y} \longrightarrow \boldsymbol{Z}$$

the pair (f_1, f_2) **locally left** (resp. **right**) **deformable** if (42.3) there exist *left* (resp. *right*) *deformation retracts* $\boldsymbol{X}_0 \subset \boldsymbol{X}$ and $\boldsymbol{Y}_0 \subset \boldsymbol{Y}$ (8.3) such that

(i) f_1 is *homotopical* on $\boldsymbol{X}_0$ (and hence f_1 is left (resp. right) deformable),
(ii) f_2 is *homotopical* on $\boldsymbol{Y}_0$ (and hence f_2 is left (resp. right) deformable), and
(iii) $f_2 f_1$ is *homotopical* on $\boldsymbol{X}_0$ (and hence $f_2 f_1$ is left (resp. right) deformable),

and call such a pair **left** (resp. **right**) **deformable** if (42.3)

(iv) f_1 sends all of $\boldsymbol{X}_0$ into $\boldsymbol{Y}_0$ (which together with (i) and (ii) implies (iii)).

We then note that (42.3(v)) such left (resp. right) deformability is equivalent to the requirement that

(v) there exist left (resp. right) deformations (r_1, s_1) on $\boldsymbol{X}$ and (r_2, s_2) on $\boldsymbol{Y}$ such that f_1 and f_2 are homotopical on the full subcategories spanned by the images of r_1 and r_2 respectively and the natural transformation

$$f_2 s_2 f_1 r_1 \colon f_2 r_2 f_1 r_1 \longrightarrow f_2 f_1 r_1 \quad (\text{resp. } f_2 s_2 f_1 r_1 \colon f_2 f_1 r_1 \longrightarrow f_2 r_2 f_1 r_1)$$

is a natural *weak equivalence*

as in that case f_2 is homotopical on the full subcategory of $\boldsymbol{Y}$ spanned by the images of r_2 and $f_1 r_1$.

This last result implies that the behavior of approximations of Quillen functors under composition (15.5) is a special case of 42.4, which is proven exactly as in 15.5, and which states that

(vi) *if, for two composable functors*

$$f_1\colon \boldsymbol{X} \longrightarrow \boldsymbol{Y} \quad \text{and} \quad f_2\colon \boldsymbol{Y} \longrightarrow \boldsymbol{Z}$$

the pair (f_1, f_2) is left (resp. right) deformable, then their composition is left (resp. right) deformable and the compositions (as defined in 15.4) of their left (resp. right) approximations are left (resp. right) approximations of their composition.

Moreover, one can often verify the deformability of two composable functors, by noting that (42.5)

(vii) *if, for two composable deformable adjunctions* (18.2)

$$f_1\colon \boldsymbol{X} \longleftrightarrow \boldsymbol{Y} \colon f_1' \quad \text{and} \quad f_2\colon \boldsymbol{Y} \longleftrightarrow \boldsymbol{Z} \colon f_2'$$

between saturated (8.4) homotopical categories, the pair (f_1, f_2) is locally left deformable and the pair (f_2', f_1') is locally right deformable, then the pair (f_1, f_2) is left deformable iff the pair (f_2', f_1') is right deformable.

Next we note that, just like Quillen adjunctions, deformable adjunctions (18.2) give rise to

18.5. Partial adjunctions and derived adjunctions. The existence of the induced partial adjunction of a Quillen functor (16.1) is a special case of 43.2 which states in part that

(i) *given a deformable adjunction* (18.2) $f\colon \boldsymbol{X} \leftrightarrow \boldsymbol{Y} :f'$, *a left deformation retract* $\boldsymbol{X}_0 \subset \boldsymbol{X}$ *on which f is homotopical and a right deformation retract* $\boldsymbol{Y}_0 \subset \boldsymbol{Y}$ *on which f' is homotopical, its adjunction induces a* **partial adjunction** *which to every pair of objects $X \in \boldsymbol{X}_0$ and $Y \in \boldsymbol{Y}_0$ assigns the unique* **partial adjunction isomorphism** (8.4)

$$\operatorname{Ho}\boldsymbol{Y}(fX, Y) \approx \operatorname{Ho}\boldsymbol{X}(X, f'Y)$$

which

(ii) *is natural in X and Y, and*

(iii) *is compatible with the given adjunction in the sense that the diagram*

$$\begin{array}{ccc} \boldsymbol{Y}(fX, Y) & \approx & \boldsymbol{X}(X, f'Y) \\ \downarrow & & \downarrow \\ \operatorname{Ho}\boldsymbol{Y}(fX, Y) & \approx & \operatorname{Ho}\boldsymbol{X}(X, f'Y) \end{array}$$

in which the top map is the given adjunction isomorphism and the vertical maps are induced by the localization functors (8.4), *commutes.*

Using these partial adjunctions and the observation that (44.1)

(iv) *the results of* 16.2 *remain valid if one replaces everywhere the Quillen adjunction* $f\colon \boldsymbol{M} \leftrightarrow \boldsymbol{N} :f'$ *by a deformable adjunction* $f\colon \boldsymbol{X} \leftrightarrow \boldsymbol{Y} :f'$ (18.2)

one then can show, using the same arguments as were used in proving 16.3 and 16.5 that (44.2)

(v) *the results of* 16.3 *remain valid if one replaces everywhere the Quillen adjunction* $f\colon \boldsymbol{M} \leftrightarrow \boldsymbol{N} :f'$ *by a deformable adjunction* $f\colon \boldsymbol{X} \leftrightarrow \boldsymbol{Y} :f'$ *and the deformation retracts* $\boldsymbol{M}_c$ *and* $\boldsymbol{N}_f$ *by a left deformation retract* $\boldsymbol{X}_0$ *of* $\boldsymbol{X}$ *on which f is homotopical and a right deformation retract* $\boldsymbol{Y}_0$ *of* $\boldsymbol{Y}$ *on which f' is homotopical*

and that (44.4)

(vi) *the results of* 16.5 *remain valid if one replaces the composable Quillen adjunctions*

$$f_1\colon \boldsymbol{M} \longleftrightarrow \boldsymbol{N} :f'_1 \quad \text{and} \quad f_2\colon \boldsymbol{N} \longleftrightarrow \boldsymbol{P} :f'_2$$

by composable deformable adjunctions

$$f_1\colon \boldsymbol{X} \longleftrightarrow \boldsymbol{Y} :f'_1 \quad \text{and} \quad f_2\colon \boldsymbol{Y} \longleftrightarrow \boldsymbol{Z} :f'_2$$

for which the pairs (f_1, f_2) and (f'_1, f'_2) are respectively left and right deformable (18.4).

We end with a comment on

18.6. The Quillen condition. A deformable adjunction (18.2) $f\colon \boldsymbol{X} \leftrightarrow \boldsymbol{Y} : f'$ is (45.1) said to satisfy the **Quillen condition** if

(i) there exist a left deformation retract $\boldsymbol{X}_0 \subset \boldsymbol{X}$ on which f is homotopical and a right deformation retract $\boldsymbol{Y}_0 \subset \boldsymbol{Y}$ on which f' is homotopical, such that, for every pair of objects $X \in \boldsymbol{X}_0$ and $Y \in \boldsymbol{Y}_0$, a map $fX \to Y \in \boldsymbol{Y}$ is a weak equivalence *iff* its adjunct $X \to f'Y \in \boldsymbol{X}$ is so,

which is readily verified to be equivalent to the requirement that

(ii) for every left deformation retract $\boldsymbol{X}_0 \subset \boldsymbol{X}$ on which f is homotopical and every right deformation retract $\boldsymbol{Y}_0 \subset \boldsymbol{Y}$ on which f' is homotopical, for every pair of objects $X \in \boldsymbol{X}_0$ and $Y \in \boldsymbol{Y}_0$, a map $fX \to Y \in \boldsymbol{Y}$ is a weak equivalence *iff* its adjunct $X \to f'Y \in \boldsymbol{X}$ is so.

The results of 17.1 and 17.3 for Quillen adjunctions then are a special case of the essentially identical results for deformable adjunctions obtained in the last sentence of 45.3 (where we assume that the categories involved are *saturated* (8.4)).

CHAPTER IV

Homotopical Cocompleteness and Completeness of Model Categories

19. Introduction

19.1. Summary. In this chapter we consider the "homotopically correct" *homotopy colimit* and *limit functors* on a model category, i.e. the homotopy colimit and limit functor which, for all diagrams indexed by a small category (and not merely the objectwise cofibrant or fibrant ones), yield the "correct" homotopy type, and show in particular that every model category M not only (by definition) is cocomplete and complete; but also is *homotopically cocomplete* and *complete* in a sense which is considerably stronger than the requirement that there exist homotopy D-colimit and D-limit functors on M for every small category D.

In more detail (in the colimit case only):

19.2. Homotopy colimit functors. Let **cat** denote the category of *small categories* (8.1). Given a model category M we then define, for every object $D \in$ **cat**,

(i) a **homotopy D-colimit functor** on M as a *left approximation* (18.3) of an arbitrary but fixed *left adjoint*
$$\mathrm{colim}^D \colon M^D \longrightarrow M$$
of the constant diagram functor $M \to M^D$,

and more generally, for every map $u \colon A \to B \in$ **cat**,

(ii) a **homotopy u-colimit functor** on M as a *left approximation* of an arbitrary but fixed *left adjoint*
$$\mathrm{colim}^u \colon M^A \longrightarrow M^B$$
of the induced diagram functor
$$u^* = M^u \colon M^B \longrightarrow M^A$$

and show that

(iii) such homotopy u-colimit functors *exist* and are *homotopically unique* (13.2),

(iv) every such homotopy u-colimit functor (k, a) on M comes with a *derived adjunction* (16.3 and 18.5(v))
$$\mathrm{Ho}\, k \colon \mathrm{Ho}\, X^A \longleftrightarrow X^B \colon \mathrm{Ho}\, u^*$$

and that

(v) these homotopy colimit functors on M are *composable*.

We also note that

(vi) *left Quillen functors are **homotopically compatible** with homotopy colimit functors in the sense that, for every left Quillen functor $f\colon M \to N$ and map $u\colon A \to B \in \mathbf{cat}$*

(vi)′ *the compositions (15.4 and 18.4) of a homotopy u-colimit functor on M with a left approximation (18.3) of the functor $f^B\colon M^B \to N^B$, and*

(vi)″ *the compositions of a left approximation of the functor $f^A\colon M^A \to N^A$ with a homotopy u-colimit functor on N,*

are respectively left approximations of the (see 19.6(v) below) canonically (naturally) isomorphic compositions

$$M^A \xrightarrow{\operatorname{colim}^u} M^B \xrightarrow{f^B} N^B \quad \text{and} \quad M^A \xrightarrow{f^A} N^A \xrightarrow{\operatorname{colim}^u} N^B$$

and hence (13.4) canonically weakly equivalent (13.2).

19.3. Homotopical cocompleteness. We show that every model category M is **homotopically cocomplete** in a sense which is stronger than the requirement that there exist homotopy D-colimit functors on M for every object $D \in \mathbf{cat}$ (19.2) or even homotopy u-colimit functors on M for every map $u\colon A \to B \in \mathbf{cat}$, as it requires the existence of what we will call a **homotopy colimit system** on M, i.e. the existence of a function which assigns

(i) to every map $u\colon A \to B \in \mathbf{cat}$, a *homotopy u-colimit functor* (k_u, a_u) on M, and

(ii) to every composable pair of maps $u\colon A \to B$ and $v\colon B \to D \in \mathbf{cat}$, a *weak equivalence* from the composition of the chosen homotopy u- and v-colimit functors to the chosen homotopy vu-colimit functor, which is *associative*

We also note that

(iii) such homotopy colimit systems on M are *homotopically unique* (13.2).

19.4. The proofs. The results mentioned in 19.2 and 19.3 will be obtained by combining some of the statements in §18 (the proofs of which are in Part II) with the **deformability result** (20.5) that

(i) *there exists, for every model category M and object $D \in \mathbf{cat}$ (19.2), a left deformation retract $M^D_{vc} \subset M^D$ (8.3) such that*

(ii) *for every map $u\colon A \to B \in \mathbf{cat}$, the functor $\operatorname{colim}^u\colon M^A \to M^B$ sends objectwise weak equivalences in M^A_{vc} to objectwise weak equivalences in M^B_c (10.2), and*

(iii) *for every left Quillen functor $f\colon M \to N$ and object $D \in \mathbf{cat}$, the functor $f^D\colon M^D \to N^D$ sends objectwise weak equivalences in M^D_{vc} to objectwise weak equivalences in N^D_{vc}.*

The proof of this result relies heavily on the fact that, for every object $D \in \mathbf{cat}$, the *category of simplices* $\mathbf{\Delta} D$ of D and its opposite $\mathbf{\Delta}^{\mathrm{op}} D$ have the property that, for every model category M, the model structure of M induces a *Reedy model structure* on the diagram categories $M^{\mathbf{\Delta} D}$ and $M^{\mathbf{\Delta}^{\mathrm{op}} D}$.

19.5. Organization of the chapter. After fixing some notation and terminology involving *colimit* and *limit functors* (in the remainder of this section) we devote two sections to *homotopy colimit* and *limit functors* (§20) and *homotopical cocompleteness* and *completeness* (§21) and then deal in the last two sections with

Reedy model structures (§22) and a proof of the *deformability result* (§23). In the last section (§24) we note that the main results of this chapter, i.e. those obtained in §20 and §21 are special cases of more general results on homotopical categories which we discuss in chapter VIII.

It thus remains to discuss

19.6. Colimit and limit functors. Given a category X and a category D,
  (i) a **D-colimit** (resp. **D-limit**) **functor** on X will be a left (resp. a right) adjoint $X^D \to X$ of the **constant diagram functor** $c^* \colon D \to X^D$ (which sends every object of X to the corresponding constant D-diagram) and, if such adjoints exist, we denote an *arbitrary but fixed* such adjoint by
$$\operatorname{colim}^D \qquad (\text{resp. } \lim^D)$$
and more generally, given a functor $u \colon A \to B$
  (ii) a **u-colimit** (resp. a **u-limit**) **functor** on X will be a left (resp. a right) adjoint $X^A \to X^B$ of the **induced diagram functor**
$$X^u = u^* \colon X^B \longrightarrow X^A$$
(which sends every functor $B \to X$ to its composition with u) and, if such adjoints exist, we denote an *arbitrary but fixed* such adjoint by
$$\operatorname{colim}^u \qquad (\text{resp. } \lim^u)$$

Similarly
  (iii) if, for two composable functors $u \colon A \to B$ and $v \colon B \to D$ there exist u-colimit (resp. u-limit) functors and v-colimit (resp. v-limit) functors (and hence vu-colimit (resp. vu-limit) functors) on X we denote by
$$\operatorname{colim}^{(v,u)} \colon \operatorname{colim}^v \operatorname{colim}^u \longrightarrow \operatorname{colim}^{vu} \qquad (\text{resp. } \lim^{(v,u)} \colon \lim^{vu} \longrightarrow \lim^v \lim^u)$$
the conjugate (see 39.10) of the identity natural transformation of the functor
$$(vu)^* = u^* v^* \colon X^D \longrightarrow X^A \ .$$

These definitions readily imply that colimit and limit functors are *composable*, i.e. that
  (iv) *for every composable pair of functors $u \colon A \to B$ and $v \colon B \to D$, every composition of a u-colimit (resp. u-limit) functor on X with a v-colimit (resp. v-limit) functor on X, if these both exist, is a vu-colimit (resp. vu-limit) functor on X,*

and that *left* and *right* adjoints are **compatible** with respectively colimit and limit functors, in the sense that
  (v) *for every adjunction $f \colon X \leftrightarrow Y : g$ and functor $u \colon A \to B$ and every pair of u-colimit (resp. u-limit) functors s on X and t on Y, the compositions*
$$f^B s \text{ and } tf^A \colon X^A \longrightarrow Y^B \qquad (\text{resp. } g^B t \text{ and } sg^A \colon Y^A \longrightarrow X^B)$$
are left (resp. right) adjoints of the composition
$$X^u g^B = g^A Y^u \colon Y^B \longrightarrow X^A \qquad (\text{resp. } Y^u f^B = f^A X^u \colon X^B \longrightarrow Y^A)$$
and hence canonically naturally isomorphic (13.1).

We end with some comments on the related notions of

19.7. Cocompleteness and completeness. One calls a category $\boldsymbol{X}$ **cocomplete** (resp. **complete**) if

(i) for every object $\boldsymbol{D} \in \mathbf{cat}$ (the category of small categories (8.1)), there exist $\boldsymbol{D}$-colimit (resp. $\boldsymbol{D}$-limit) functors on $\boldsymbol{X}$ (19.6(i)),

and such cocompleteness (resp. completeness) implies that

(ii) *for every map* $u\colon \boldsymbol{A} \to \boldsymbol{B} \in \mathbf{cat}$, *there exist u-colimit (resp. u-limit) functors on $\boldsymbol{X}$* (19.6(ii)),

for instance the functor which sends an object $T \in \boldsymbol{X}^{\boldsymbol{A}}$ to the functor $\boldsymbol{B} \to \boldsymbol{X}$ which associates with each object $B \in \boldsymbol{B}$ the object (19.6)

$$\operatorname{colim}^{(u\downarrow B)} j^* T \in \boldsymbol{X} \qquad (\text{resp. } \lim^{(B\downarrow u)} j^* T \in \boldsymbol{X})$$

where j denotes the forgetful functor

$$j\colon (u\downarrow B) \longrightarrow \boldsymbol{A} \qquad (\text{resp. } j\colon (B\downarrow u) \longrightarrow \boldsymbol{A}) \ .$$

However, in spite of 19.6(iv), it is in general *not* possible to find for each map $u \in \mathbf{cat}$ a u-colimit (resp. a u-limit) functor on $\boldsymbol{X}$ such that these functors form a *diagram* indexed by the category $\mathbf{cat}$, but only what we will call a **colimit** (resp. a **limit**) **system** on $\boldsymbol{X}$, and by which we mean a function F which assigns

(iii) to every object $\boldsymbol{D} \in \mathbf{cat}$ the *diagram category* $\boldsymbol{X}^{\boldsymbol{D}}$,

(iv) to every map $u\colon \boldsymbol{A} \to \boldsymbol{B} \in \mathbf{cat}$, a *$u$-colimit (resp. u-limit) functor* $Fu\colon F\boldsymbol{A} \to F\boldsymbol{B}$, and

(v) to every composable pair of maps $u\colon \boldsymbol{A} \to \boldsymbol{B}$ and $v\colon \boldsymbol{B} \to \boldsymbol{D} \in \mathbf{cat}$ the *conjugate* (see 39.10)

$$F(v,u)\colon (Fv)(Fu) \longrightarrow F(vu) \qquad (\text{resp. } F(v,u)\colon F(vu) \longrightarrow (Fv)(Fu))$$

of the identity natural transformation

$$(vu)^* = u^* v^* \colon F\boldsymbol{D} \longrightarrow F\boldsymbol{A} \ .$$

Clearly

(vi) *such a colimit (resp. limit) system on $\boldsymbol{X}$ exists iff $\boldsymbol{X}$ is cocomplete (resp. complete).*

In fact in that case a for our purposes convenient example of such a colimit (resp. limit) system on $\boldsymbol{X}$ is the function

$$\operatorname{colim}^{(\mathbf{cat})} \qquad (\text{resp. } \lim^{(\mathbf{cat})})$$

which assigns to every map $u\colon \boldsymbol{A} \to \boldsymbol{B} \in \mathbf{cat}$ the u-colimit (resp. u-limit) functor (19.6(ii))

$$\operatorname{colim}^{(\mathbf{cat})} u = \operatorname{colim}^u \qquad (\text{resp. } \lim^{(\mathbf{cat})} u = \lim^u)$$

and to every composable pair of maps $u\colon \boldsymbol{A} \to \boldsymbol{B}$ and $v\colon \boldsymbol{B} \to \boldsymbol{D} \in \mathbf{cat}$, the natural isomorphism (19.6(iii))

$$\operatorname{colim}^{(\mathbf{cat})}(v,u) = \operatorname{colim}^{(v,u)} \qquad (\text{resp. } \lim^{(\mathbf{cat})}(v,u) = \lim^{(v,u)}) \ .$$

20. Homotopy colimit and limit functors

Given a model category M, we now

(i) define, for every functor $u\colon A \to B$ between small categories (8.1) *homotopy u-colimit* and *u-limit functors* on M,

(ii) prove their existence, homotopical uniqueness (13.2), composability and homotopical compatibility with Quillen functors, and

(iii) note that they give rise to derived adjunctions which are compatible with composition.

We thus start with defining

20.1. Homotopy colimit and limit functors. Given a model category M and an object $D \in \mathbf{cat}$ (the category of small categories (8.1)),

(i) a **homotopy D-colimit** (resp. **D-limit**) **functor** on M will be a left (resp. a right) approximation (18.3) of the functor $\operatorname{colim}^D\colon M^D \to M$ (resp. $\lim^D\colon M^D \to M$) (19.6(i)),

and more generally, given a map $u\colon A \to B \in \mathbf{cat}$,

(ii) a **homotopy u-colimit** (resp. **u-limit**) **functor** on M will be a left (resp. a right) approximation of the functor $\operatorname{colim}^u\colon M^A \to M^B$ (resp. $\lim^u\colon M^A \to M^B$) (19.6(ii))

and hence (18.3(i))

(iii) *such homotopy colimit (resp. limit) functors on M, if they exist, are homotopically unique (13.2).*

Next we note

20.2. Existence of homotopy colimit and limit functors and their derived adjunctions. *For every model category M and map $u\colon A \to B \in \mathbf{cat}$*

(i) *there exist homotopy u-colimit (resp. u-limit) functors on M, and*

(ii) *every homotopy u-colimit (resp. u-limit) functor (k,a) on M comes with a **derived adjunction***

$$\operatorname{Ho} k\colon \operatorname{Ho} M^A \longleftrightarrow \operatorname{Ho} M^B :\!\operatorname{Ho} u^* \qquad (\textit{resp. } \operatorname{Ho} u^*\colon \operatorname{Ho} M^B \longleftrightarrow \operatorname{Ho} M^A :\!\operatorname{Ho} k)$$

associated (16.3 and 18.5(v)) with the pair of approximations

$$\bigl((k,a),(u^*,1_{u_*})\bigr) \qquad (\textit{resp. } \bigl((u^*,1_{u^*}),(k,a)\bigr))$$

which has as its counit (resp. unit) the natural transformation

$$\operatorname{Ho}\bigl(e(au^*)\bigr)\colon (\operatorname{Ho} k)(\operatorname{Ho} u^*) \longrightarrow 1_{\operatorname{Ho} M^B}$$
$$(\textit{resp. } \operatorname{Ho}\bigl((au^*)e\bigr)\colon 1_{\operatorname{Ho} M^B} \longrightarrow (\operatorname{Ho} k)(\operatorname{Ho} u^*))$$

where

$$e\colon \operatorname{colim}^u u^* \longrightarrow 1_{M^B} \qquad (\textit{resp. } c\colon 1_{M^B} \longrightarrow \lim^u u^*)$$

denotes the counit (resp. the unit) of the adjunction

$$\operatorname{colim}^u\colon M^A \longleftrightarrow M^B :\!u^* \qquad (\textit{resp. } u^*\colon M^B \longleftrightarrow M^A :\!\lim^u) \ .$$

20.3. Composability of homotopy colimit and limit functors and their derived adjunctions. *For every model category M, composable pair of maps $u\colon A \to B$ and $v\colon B \to D \in$ cat and homotopy u-colimit (resp. u-limit) functor (k_u, a_u) and homotopy v-colimit (resp. v-limit) functor (k_v, a_v) on M,*

(i) *their* **"composition"**

$$(k_v k_u, \operatorname{colim}^{(v,u)} a_v a_u) \qquad (\text{resp. } (k_v k_u, a_v a_u \lim{}^{(v,u)}))$$

in which $\operatorname{colim}^{(v,u)}$ (resp. $\lim^{(v,u)}$) is as in 19.6(ii) and $a_v a_u$ denotes the diagonal of the commutative diagram

$$\begin{array}{ccc} k_v k_u & \xrightarrow{a_v} & \operatorname{colim}^v k_u \\ a_u \downarrow & & \downarrow a_u \\ k_v \operatorname{colim}^u & \xrightarrow{a_v} & \operatorname{colim}^v \operatorname{colim}^u \end{array} \qquad (\text{resp. } \begin{array}{ccc} \lim^v \lim^u & \xrightarrow{a_v} & k_v \lim^u \\ a_u \downarrow & & \downarrow a_u \\ \lim^v k_u & \xrightarrow{a_v} & k_v k_u \end{array})$$

is a homotopy vu-colimit (resp. vu-limit) functor on M, and

(ii) *the composition of their derived adjunctions (20.2) is the derived adjunction of their "composition".*

We also note the

20.4. Homotopical compatibility of homotopy colimit and limit functors with Quillen functors. *For every Quillen adjunction $f\colon M \leftrightarrow N : g$ (14.1) and map $u\colon A \to B \in$ cat, the left Quillen functor f is **homotopically compatible** with the homotopy u-colimit functors on M and N and right Quillen functor g is **homotopically compatible** with the homotopy u-limit functors on M and N in the sense that*

(i) *the compositions of a homotopy u-colimit functor on M with a left approximation (18.3) of f^B and of a left approximation of f^A with a homotopy u-colimit functor on N are respectively left approximations of the canonically (naturally) isomorphic (19.6(v)) compositions*

$$M^A \xrightarrow{\operatorname{colim}^u} M^B \xrightarrow{f^B} N^B \qquad \text{and} \qquad M^A \xrightarrow{f^A} N^A \xrightarrow{\operatorname{colim}^u} N^B$$

and hence are all canonically weakly equivalent (13.4), and dually

(ii) *the compositions of a homotopy u-limit functor on N with a right approximation of g^B and of a right approximation of g^A with a homotopy u-limit functor on M are respectively right approximations of the canonically (naturally) isomorphic compositions*

$$N^A \xrightarrow{\lim^u} N^B \xrightarrow{g^B} M^B \qquad \text{and} \qquad N^A \xrightarrow{g^A} M^A \xrightarrow{\lim^u} M^B$$

and hence are all canonically weakly equivalent.

The proof of these three propositions will use the following result which we will prove in 23.4.

20.5. Deformability result for colimit and limit functors. *There exist for every model category M and object $D \in $ cat, a full subcategory*

$$(M^D)_{vc} \subset M^D \qquad (\text{resp. } (M^D)_{vf} \subset M^D)$$

such that

  (i) *$(M^D)_{vc}$ (resp. $(M^D)_{vf}$) is a left (resp. a right) deformation retract (8.3) of M^D,*
  (ii) *for every Quillen adjunction $f \colon M \leftrightarrow N \colon g$ (14.1) and object $D \in $ cat, the functor*

$$f^D \colon M^D \longrightarrow N^D \qquad (\text{resp. } g^D \colon N^D \longrightarrow M^D)$$

  sends objectwise weak equivalences in $(M^D)_{vc}$ (resp. $(N^D)_{vf}$) to objectwise weak equivalences in $(N^D)_{vc}$ (resp. $(M^D)_{vf}$), and
  (iii) *for every model category M and map $u \colon A \to B \in $ cat, the functor (19.6)*

$$\operatorname{colim}^u \colon M^A \longrightarrow M^B \qquad (\text{resp. } \lim^u \colon M^A \longrightarrow M^B)$$

  sends objectwise weak equivalences in $(M^A)_{vc}$ (resp. $(M^A)_{vf}$) to objectwise weak equivalences in M^B_c (resp. M^B_f) (10.2).

20.6. Proofs (of the colimit halves) of 20.2, 20.3 and 20.4. The first part of 20.2 follows from 18.3(ii) and the fact that, in view of 20.5(i) and (iii), the functor colim^u is left deformable (18.1) and the first part of 20.3 is a consequence of 18.4(vi) and (vii), the saturation of M and hence of M^A and M^B (8.4 and 10.2), the right deformability (18.4) of the pair (M^v, M^u) and, in view of 20.5(iii), the local left deformability of the pair $(\operatorname{colim}^u, \operatorname{colim}^v)$. The second part of 20.2 follows from the observation that if (r, s) is a left deformation (8.2) of M^A into M^A_{vc} (20.5), then the resulting commutative diagram

$$\begin{array}{ccccc}
(kr)u^* & \xrightarrow{(ks)u^*} & ku^* & & \\
{\scriptstyle (ar)u^*}\downarrow & & {\scriptstyle au^*}\downarrow & \searrow^{e(au^*)} & \\
(\operatorname{colim}^u r)u^* & \xrightarrow{(\operatorname{colim}^u s)u^*} & \operatorname{colim}^u u^* & \xrightarrow{e} & 1_{M^B}
\end{array}$$

gives rise to the commutative diagram

$$\begin{array}{ccc}
(\operatorname{Ho} kr)(\operatorname{Ho} u^*) & \longrightarrow & (\operatorname{Ho} k)(\operatorname{Ho} u^*) \\
\downarrow & & \downarrow {\scriptstyle \operatorname{Ho}(e(au^*))} \\
(\operatorname{Ho} \operatorname{colim}^u r)(\operatorname{Ho} u^*) & \longrightarrow & 1_{\operatorname{Ho} M^B}
\end{array}$$

and the observation that, in view of 16.3 and 18.5(v), the natural transformation $(\operatorname{Ho} k)(\operatorname{Ho} u^*) \to 1_{\operatorname{Ho} M^B}$ obtained by going counter clockwise around this diagram is exactly the desired counit.

To prove the second part of 20.3 one notes that in the commutative diagram

$$
\begin{array}{ccccccccc}
k_v k_u u^* v^* & \xrightarrow{a_u} & k_v \operatorname{colim}^u u^* v^* & \xrightarrow{e_u} & k_v v^* & \xrightarrow{a_v} & \operatorname{colim}^v v^* & \xrightarrow{e_v} & 1_{M^B} \\
\downarrow{1} & & \downarrow{a_v} & & \downarrow{a_v} & & & & \downarrow{1} \\
 & & \operatorname{colim}^v \operatorname{colim}^u u^* v^* & \xrightarrow{e_u} & \operatorname{colim}^v v^* & \xrightarrow{\quad e_v \quad} & & & 1_{M^B} \\
 & & \downarrow{\operatorname{colim}^{(v,u)}} & & & & & & \downarrow{1} \\
k_v k_u (vu)^* & \longrightarrow & \operatorname{colim}^{vu} (vu)^* & \xrightarrow{\quad\quad e_{vu} \quad\quad} & & & & & 1_{M^B}
\end{array}
$$

in which e_u, e_v and e_{vu} are (induced by) the relevant counits, the top row, in view of 16.5 and 18.5(vi), induces the counit of the adjunction associated with the composition, in the sense of 15.4, of (k_u, a_u) and (k_v, a_v), while the bottom row, in view of the first half of the proposition, induces the counit of the adjunction associated with their "composition", in the sense of 20.3(i).

Finally, 20.4(i) follows from 18.4(vi) and the observation that, in view of 20.5(ii), the pairs

$$(\operatorname{colim}^u \colon \mathbf{M^A} \to \mathbf{M^B}, f^B) \quad \text{and} \quad (f^A, \operatorname{colim}^u \colon \mathbf{N^A} \to \mathbf{N^B})$$

are left deformable (18.4).

21. Homotopical cocompleteness and completeness

In this section we show that every model category $\mathbf{M}$ is *homotopically cocomplete* and *complete* in the sense that there exist *homotopy colimit* and *limit systems* on $\mathbf{M}$, which we define as a kind of left and right approximations of the colimit and limit systems $\operatorname{colim}^{(\mathbf{cat})}$ and $\lim^{(\mathbf{cat})}$ (19.7).

To do this it is convenient to first introduce the notions of

21.1. Left and right cat-systems. Given a model category $\mathbf{M}$, a **left** (resp. a **right**) **cat-system** on $\mathbf{M}$ will be a function F which assigns

  (i) to every object $\mathbf{D} \in \mathbf{cat}$ the diagram category $F\mathbf{D} = \mathbf{M^D}$,
  (ii) to every map $u\colon \mathbf{A} \to \mathbf{B} \in \mathbf{cat}$ a (not necessarily homotopical) functor $fu\colon F\mathbf{A} \to F\mathbf{B}$, and
  (iii) to every composable pair of maps $u\colon \mathbf{A} \to \mathbf{B}$ and $v\colon \mathbf{B} \to \mathbf{D} \in \mathbf{cat}$, a natural *weak equivalence* (called **composer**)

$$F(v,u)\colon (Fv)(Fu) \longrightarrow F(vu) \quad (\text{resp. } F(v,u)\colon F(vu) \longrightarrow (Fv)(Fu))$$

which is *associative* in the sense that

  (iv) for every three composable maps $u\colon \mathbf{A} \to \mathbf{B}$, $v\colon \mathbf{B} \to \mathbf{D}$ and $x\colon \mathbf{D} \to \mathbf{E} \in \mathbf{cat}$, the following diagram commutes

$$
\left(
\begin{array}{ccc}
(Fx)(Fv)(Fu) & \xrightarrow{(Fx)F(v,u)} & (Fx)F(vu) \\
\downarrow{F(x,v)(Fu)} & & \downarrow{F(x,vu)} \\
F(xv)(Fu) & \xrightarrow{F(xv,u)} & F(xvu)
\end{array}
\quad (\text{resp.} \quad
\begin{array}{ccc}
(Fx)F(vu) & \xrightarrow{(Fx)F(v,u)} & (Fx)(Fv)(Fu) \\
\uparrow{F(x,vu)} & & \uparrow{F(x,v)(Fu)} \\
F(xvu) & \xrightarrow{F(xv,u)} & F(xv)(Fu)
\end{array}
\right)
$$

and such a **cat**-system will be called **homotopical** if, for every map $u\colon \boldsymbol{A} \to \boldsymbol{B} \in$ **cat**, the functor $Fu\colon F\boldsymbol{A} \to F\boldsymbol{B}$ is *homotopical*.

Furthermore a **map** $h\colon F \to G$ between two such left (resp. right) **cat**-systems F and G will be a function h which assigns to every map $u\colon \boldsymbol{A} \to \boldsymbol{B} \in$ **cat** a natural transformation $hu\colon Fu \to Gu$, which *commutes with the composers* in the sense that

(v) for every composable pair of maps $u\colon \boldsymbol{A} \to \boldsymbol{B}$ and $v\colon \boldsymbol{B} \to \boldsymbol{D} \in$ **cat**, the following diagram commutes:

$$\begin{array}{ccc} (Fv)(Fu) \xrightarrow{F(v,u)} F(vu) & & F(vu) \xrightarrow{F(v,u)} (Fv)(Fu) \\ {\scriptstyle (hv)(hu)}\Big\downarrow \quad {\scriptstyle h(vu)}\Big\downarrow & (\text{resp.} & {\scriptstyle h(vu)}\Big\downarrow \quad {\scriptstyle (hv)(hu)}\Big\downarrow \quad) \\ (Gv)(Gu) \xrightarrow[G(v,u)]{} G(vu) & & G(vu) \xrightarrow[G(v,u)]{} (Gv)(Gu) \end{array}$$

and such a map will be called a **weak equivalence** whenever, for every map $u \in$ **cat**, the natural transformation hu is a *natural weak equivalence*.

We denote the resulting homotopical category of left (resp. right) **cat**-systems by

$$\textbf{cat}_L\textbf{-syst} \qquad (\text{resp. } \textbf{cat}_R\textbf{-syst})$$

and its full subcategory spanned by the homotopical **cat**-systems (iv) by

$$\textbf{cat}_L\textbf{-syst}_\text{w} \qquad (\text{resp. } \textbf{cat}_R\textbf{-syst}_\text{w}) \ .$$

21.2. Example. For every model category $\boldsymbol{M}$, the colimit (resp. the limit) systems on $\boldsymbol{M}$ (19.7) are *left* (resp. *right*) **cat***-systems* on $\boldsymbol{M}$, and so is in particular the colimit system $\mathrm{colim}^{(\mathbf{cat})}$ (resp. the limit system $\lim^{(\mathbf{cat})}$) (19.7).

Now we can define

21.3. Homotopy colimit and limit systems. Given a model category $\boldsymbol{M}$, a **homotopy colimit** (resp. **limit**) **system** on $\boldsymbol{M}$ will be a pair (K,a) consisting of an object (21.1)

$$K \in \textbf{cat}_L\textbf{-syst}_\text{w} \qquad (\text{resp. } K \in \textbf{cat}_R\textbf{-syst}_\text{w})$$

and a map (21.1 and 21.2)

$$a\colon K \longrightarrow \mathrm{colim}^{(\mathbf{cat})} \in \textbf{cat}_L\textbf{-syst} \qquad (\text{resp. } a\colon \lim^{(\mathbf{cat})} \longrightarrow K \in \textbf{cat}_R\textbf{-syst})$$

such that

(i) for every map $u \in$ **cat**, the pair (Ku, au) is a *homotopy u-colimit* (resp. *u-limit*) *functor* on $\boldsymbol{M}$ (20.1), and

(ii) (K,a) is a *homotopically terminal* (resp. *initial*) object of the "homotopical category of homotopical left (resp. right) **cat**-systems on $\boldsymbol{M}$ over $\mathrm{colim}^{(\mathbf{cat})}$ (resp. under $\lim^{(\mathbf{cat})}$)

$$(\textbf{cat}_L\textbf{-syst}_\text{w} \downarrow \mathrm{colim}^{(\mathbf{cat})}) \qquad (\text{resp. } (\lim^{(\mathbf{cat})} \downarrow \textbf{cat}_R\textbf{-syst}_\text{w}))$$

which has as objects the above pairs (K,a) and, for every two such pairs (K_1, a_1) and (K_2, a_2) as maps and weak equivalences $(K_1, a_1) \to (K_2, a_2)$ the maps and weak equivalences $t\colon K_1 \to K_2$ (21.1) such that $a_2 t = a_1$ (resp. $t a_1 = a_2$).

It then follows from 13.2(ii) and 13.3(vii) that

(iii) *if one object of*

$$(\mathbf{cat}_L\text{-}\mathbf{syst}_w \downarrow \mathrm{colim}^{(\mathbf{cat})}) \qquad (\textit{resp. } (\mathrm{lim}^{(\mathbf{cat})} \downarrow \mathbf{cat}_R\text{-}\mathbf{syst}_w))$$

satisfies both (i) *and* (ii), *then every object which satisfies one of* (i) *and* (ii) *also satisfies the other*

and hence (13.3(vii))

(iv) *the homotopy colimit (resp. limit) systems on* $\boldsymbol{M}$, *if they exist, are homotopically unique* (13.2).

It thus remains to prove the promised

21.4. Homotopical cocompleteness and completeness of model categories. *Let* $\boldsymbol{M}$ *be a model category. Then*

(i) $\boldsymbol{M}$ *is* **homotopically cocomplete** *and* **complete** *in the sense that*
(ii) *there exist homotopy colimit and limit systems on* $\boldsymbol{M}$ (21.3).

Moreover 21.3*(iii)*

(iii) *every object of*

$$(\mathbf{cat}_L\text{-}\mathbf{syst}_w \downarrow \mathrm{colim}^{(\mathbf{cat})}) \qquad (\textit{resp. } (\mathrm{lim}^{(\mathbf{cat})} \downarrow \mathbf{cat}_R\text{-}\mathbf{syst}_w))$$

which satisfies either 21.3(i) *or* 21.3(ii) *satisfies both and hence is such a homotopy colimit (resp. limit) system.*

Proof (of the colimit half). Let (r, s) be a pair of functions which assign to every object $\boldsymbol{D} \in \mathbf{cat}$ a left deformation $(r_{\boldsymbol{D}}, s_{\boldsymbol{D}})$ of $\boldsymbol{M}^{\boldsymbol{D}}$ into $\boldsymbol{M}_{vc}^{\boldsymbol{D}}$ (8.3 and 20.5(i)), and for every left **cat**-system K on $\boldsymbol{M}$ (21.1), let Kr denote the function which assigns

(i) to every map $u \colon \boldsymbol{A} \to \boldsymbol{B} \in \mathbf{cat}$, the functor

$$(Kr)u = (Ku)r_{\boldsymbol{A}} \colon \boldsymbol{M}^{\boldsymbol{A}} \longrightarrow \boldsymbol{M}^{\boldsymbol{B}}$$

and

(ii) to every composable pair of maps $u \colon \boldsymbol{A} \to \boldsymbol{B}$ and $v \colon \boldsymbol{B} \to \boldsymbol{D} \in \mathbf{cat}$, the composition $(Kr)(v, u)$ of the natural transformation

$$(Kr)v(Kr)u = (Kv)r_{\boldsymbol{B}}(Ku)r_{\boldsymbol{A}} \xrightarrow{(Kv)s_{\boldsymbol{B}}(Ku)r_{\boldsymbol{A}}} (Kv)(Ku)r_{\boldsymbol{A}}$$

with the natural weak equivalence

$$(Kv)(Ku)r_{\boldsymbol{A}} \xrightarrow{K(v,u)r_{\boldsymbol{A}}} K(vu)r_{\boldsymbol{A}} = (Kr)(vu)$$

and let Ks denote the function which assigns

(iii) to every map $u \colon \boldsymbol{A} \to \boldsymbol{B} \in \mathbf{cat}$, the natural transformation

$$(Ks)u = (Ku)s_{\boldsymbol{A}} \colon (Ku)r_{\boldsymbol{A}} \longrightarrow Ku \ .$$

Clearly

(iv) *if K is a homotopical left* **cat***-system* (21.1), *then so is Kr and the function Ks is a weak equivalence* $Kr \to K \in \mathbf{cat}_L\text{-}\mathbf{syst}_w$ (21.1).

Furthermore the saturation of $\boldsymbol{M}$, together with 18.3(iii), 18.4(vii) and 20.5(iii) and the fact that, for every map $u \colon \boldsymbol{A} \to \boldsymbol{B} \in \mathbf{cat}$, the functor $\boldsymbol{M}^u \colon \boldsymbol{M}^{\boldsymbol{B}} \to \boldsymbol{M}^{\boldsymbol{A}}$ is homotopical, implies that

(v) $\mathrm{colim}^{(\mathbf{cat})} r$ *is a homotopical left* **cat***-system and the function* $\mathrm{colim}^{(\mathbf{cat})} s$ *is a map* $\mathrm{colim}^{(\mathbf{cat})} r \to \mathrm{colim}^{(\mathbf{cat})} \in \mathbf{cat}_L\text{-}\mathbf{syst}$.

In view of the *naturality in K* of the functions Kr and Ks, it therefore follows that

(vi) *every object $(K, a) \in (\mathbf{cat}_L\text{-}\mathbf{syst}_w \downarrow \mathrm{colim}^{(\mathbf{cat})})$ (21.2 and 21.3) gives rise to a zigzag in $(\mathbf{cat}_L\text{-}\mathbf{syst}_w \downarrow \mathrm{colim}^{(\mathbf{cat})})$*

$$(K, a) \xleftarrow{Ks} (Kr, as) \xrightarrow{ar} (\mathrm{colim}^{(\mathbf{cat})} r, \mathrm{colim}^{(\mathbf{cat})} s)$$

in which as denotes the diagonal of the commutative diagram

$$\begin{array}{ccc} Kr & \xrightarrow{ar} & \mathrm{colim}^{(\mathbf{cat})} r \\ Ks \downarrow & & \downarrow \mathrm{colim}^{(\mathbf{cat})} s \\ K & \xrightarrow{a} & \mathrm{colim}^{(\mathbf{cat})} \end{array}$$

which is natural in (K, a)

and 21.4(ii) now is a consequence of 13.3, 21.3 and the observation that, as (20.5(iii)) for every map $u \colon \boldsymbol{A} \to \boldsymbol{B} \in \mathbf{cat}$ the natural transformation

$$((\mathrm{colim}^{(\mathbf{cat})} s)r)u = (\mathrm{colim}^{(\mathbf{cat})} s)ur_{\boldsymbol{A}} = \mathrm{colim}^u s_{\boldsymbol{A}} r_{\boldsymbol{A}}$$

is a natural weak equivalence, the map

$$(\mathrm{colim}^{(\mathbf{cat})} s)r \colon (\mathrm{colim}^{(\mathbf{cat})} r)r \longrightarrow \mathrm{colim}^{(\mathbf{cat})} r \in \mathbf{cat}_L\text{-}\mathbf{syst}_w$$

is a weak equivalence.

22. Reedy model categories

In preparation for the next section, where we will give a proof of 20.5, we

(i) review the notion of a *Reedy category* and, for every Reedy category $\boldsymbol{B}$ and model category $\boldsymbol{M}$, the resulting *Reedy model structure* on the diagram category $\boldsymbol{M}^{\boldsymbol{B}}$ [**Ree74**], [**Hir03**] and [**Hov99**], and

(ii) discuss briefly the rather useful special case of a Reedy category *with fibrant or cofibrant constants*, i.e. [**Hir03**] a Reedy category $\boldsymbol{B}$ with the property that, for every model category $\boldsymbol{M}$ the adjunction (19.6)

$$\mathrm{colim}^{\boldsymbol{B}} \colon \boldsymbol{M}^{\boldsymbol{B}} \longleftrightarrow \boldsymbol{M} \colon c^* \qquad \text{or} \qquad c^* \colon \boldsymbol{M} \longleftrightarrow \boldsymbol{M}^{\boldsymbol{B}} \colon \lim^{\boldsymbol{B}}$$

is a Quillen adjunction (14.1).

We thus start with

22.1. Reedy category. A **Reedy category** consists of a small category $\boldsymbol{B}$ (8.1), together with two subcategories $\overrightarrow{\boldsymbol{B}}$ and $\overleftarrow{\boldsymbol{B}}$ such that

(i) there exists a **degree function** which assigns to every object $B \in \boldsymbol{B}$ a non-negative integer $\deg B$ such that all non-identity maps of $\overrightarrow{\boldsymbol{B}}$ raise the degree and all non-identity maps of $\overleftarrow{\boldsymbol{B}}$ lower it, and

(ii) every map $b \in \boldsymbol{B}$ admits a *functorial* factorization $b = \overrightarrow{b}\, \overleftarrow{b}$ with $\overrightarrow{b} \in \overrightarrow{\boldsymbol{B}}$ and $\overleftarrow{b} \in \overleftarrow{\boldsymbol{B}}$.

Such Reedy categories give rise to

22.2. Reedy model structures. For every model category $\boldsymbol{M}$ and Reedy category $\boldsymbol{B}$ (22.1) the **Reedy model structure** on the diagram category $\boldsymbol{M}^{\boldsymbol{B}}$ will be the model structure in which the weak equivalences, the cofibrations and the fibrations (which will sometimes be referred to as **Reedy weak equivalences**, **Reedy cofibrations** and **Reedy fibrations**) are as in the last part of the following

22.3. Implicit description of the Reedy model structure. *For every model category M and Reedy category B (22.1),*

(i) *the diagram category $M^{\vec{B}}$ admits a model structure in which the weak equivalences and the fibrations are the objectwise ones,*
(ii) *the diagram category $M^{\overleftarrow{B}}$ admits a model structure in which the weak equivalences and the cofibrations are the objectwise ones, and*
(iii) *the diagram category M^{B} admits a model structure in which a map is a weak equivalence, a cofibration or a fibration iff its restrictions to both $\vec{B}$ and $\overleftarrow{B}$ are so in the sense of (i) and (ii).*

22.4. Corollary. *If $f\colon M \leftrightarrow N :g$ is a Quillen adjunction (14.1), then so is, for every Reedy category B, the induced adjunction*
$$f^{B}\colon M^{B} \longleftrightarrow N^{B} :g^{B}\ .$$

To give a more explicit description (than 22.3) of the (trivial) cofibrations and the (trivial) fibrations in these Reedy model structures it is convenient to first introduce the notions of

22.5. Latching and matching objects. Given a model category M, a Reedy category B and an object $X \in M^{B}$ one can, for every object $B \in B$ consider the *maximal* subcategories
$$\partial(\vec{B}\downarrow B) \subset (\vec{B}\downarrow B) \qquad \text{and} \qquad \partial(B\downarrow\overleftarrow{B}) \subset (B\downarrow\overleftarrow{B})$$
which do *not* contain the identity maps 1_B of B and define the **latching object** $\mathrm{L}XB$ and the **matching object** $\mathrm{M}XB$ of X at B as the object (19.6)
$$\mathrm{L}XB = \operatorname{colim}^{\partial(\vec{B}\downarrow B)} j^{*}X \qquad \text{and} \qquad \mathrm{M}XB = \lim{}^{\partial(B\downarrow\overleftarrow{B})} j^{*}X$$
where j denotes the forgetful functor.

In terms of these latching and matching objects one then can formulate the following

22.6. Explicit description of the Reedy model structure. *Let M be a model category and B a Reedy category (22.1). In the Reedy model structure on M^{B} (22.3) a map $f\colon X \to Y \in M^{B}$*

(i) *is a weak equivalence iff it is an objectwise weak equivalence,*
(ii) *is a (trivial) cofibration if, for every object $B \in B$, the induced map (22.5)*
$$XB \amalg_{\mathrm{L}XB} \mathrm{L}YB \longrightarrow YB \in M$$
is a (trivial) cofibration, and
(iii) *is a (trivial) fibration iff, for every object $B \in B$, the induced map*
$$XB \longrightarrow YB \amalg_{\mathrm{M}YB} \mathrm{M}XB \in M$$
is a (trivial) fibration.

For a proof of 22.3 and 22.6 we refer the reader to [**Hir03**] and [**Hov99**].

A rather useful kind of Reedy category are the so-called

22.7. Categories with fibrant or cofibrant constants. A **category with fibrant** (resp. **cofibrant**) **constants** will be a Reedy category B such that, for every model category M, the adjunction (19.6)
$$\operatorname{colim}^B\colon M^B \longleftrightarrow M \colon c^* \qquad (\text{resp. } c^*\colon M \longleftrightarrow M^B \colon \lim^B)$$
is a Quillen adjunction (14.1).

A necessary and sufficient condition for this to happen is provided by

22.8. Proposition. *A Reedy category B has fibrant (resp. cofibrant) constants (22.7) iff the subcategory $\overleftarrow{B}$ (resp. $\overrightarrow{B}$) is a disjoint union of categories with a terminal (resp. an initial) object.*

Proof (of the cofibrant half). The proof of the "if" part is straightforward and if one endows the category **set** of small sets (8.1) with the model structure (9.7(v)) in which the cofibrations are the monomorphisms and the fibrations are the epimorphisms, then the requirement that the diagram $c^*[0] \in \mathbf{set}^B$ (13.1) be Reedy cofibrant readily implies the "only if" part.

22.9. Corollary. *If M is a model category, B a Reedy category with fibrant (resp. cofibrant) constants (22.7) which has an initial (resp. a terminal) object B_0, and $X \in M^B$ is a Reedy cofibrant (resp. fibrant) diagram in which all maps are weak equivalences, then the induced map*
$$XB_0 \longrightarrow \operatorname{colim}^B X \in M \qquad (\text{resp. } \lim^B X \longrightarrow XB_0 \in M)$$
is also a weak equivalence.

Important examples of categories with fibrant or cofibrant constants are the

22.10. Categories of simplices and their opposites. For every integer $n \geq 0$ let $[n]$ denote the category which has as objects the integers $0, \ldots, n$ and which has exactly one map $i \to j$ whenever $i \leq j$. For every small category D (8.1) its **category of simplices** ΔD then will be the category which has as objects the functors $[n] \to D$ ($n \geq 0$) and which has as maps
$$(E_1\colon [n_1] \to D) \longrightarrow (E_2\colon [n_2] \to D)$$
the commutative triangles of the form

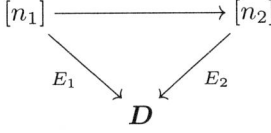

The opposite of ΔD will be denoted by $\Delta^{\mathrm{op}} D$.

Clearly (22.8)

(i) ΔD and $\Delta^{\mathrm{op}} D$ are Reedy categories with fibrant and cofibrant constants respectively (for which the subcategories
$$\overrightarrow{\Delta D} \subset \Delta D \qquad \text{and} \qquad \overleftarrow{\Delta^{\mathrm{op}} D} \subset \Delta^{\mathrm{op}} D$$

consist of the above triangles in which the horizontal map is 1-1 and the subcategories

$$\overleftarrow{\mathbf{\Delta} \mathbf{D}} \subset \mathbf{\Delta} \mathbf{D} \quad \text{and} \quad \overrightarrow{\mathbf{\Delta}^{\mathrm{op}} \mathbf{D}} \subset \mathbf{\Delta}^{\mathrm{op}} \mathbf{D}$$

consist of those in which this map is onto)

and hence *(14.4 and 22.7)*

(ii) *for every model category $\mathbf{M}$, the functor*

$$\operatorname{colim}^{\mathbf{\Delta} \mathbf{D}} \colon \mathbf{M}^{\mathbf{\Delta} \mathbf{D}} \longrightarrow \mathbf{M} \quad (\text{resp. } \lim{}^{\mathbf{\Delta}^{\mathrm{op}} \mathbf{D}} \colon \mathbf{M}^{\mathbf{\Delta}^{\mathrm{op}} \mathbf{D}} \longrightarrow \mathbf{M})$$

sends Reedy weak equivalences between Reedy cofibrant (resp. fibrant) diagrams to weak equivalences between cofibrant (resp. fibrant) objects.

Moreover one also readily verifies that

(iii) *for every functor $u \colon \mathbf{A} \to \mathbf{B}$ between small categories, the induced functor*

$$(\mathbf{\Delta} u)^* \colon \mathbf{M}^{\mathbf{\Delta} \mathbf{B}} \to \mathbf{M}^{\mathbf{\Delta} \mathbf{A}} \quad (\text{resp. } (\mathbf{\Delta}^{\mathrm{op}} u)^* \colon \mathbf{M}^{\mathbf{\Delta}^{\mathrm{op}} \mathbf{B}} \to \mathbf{M}^{\mathbf{\Delta}^{\mathrm{op}} \mathbf{A}})$$

preserves Reedy weak equivalences between Reedy cofibrant (resp. fibrant) diagrams.

We end with generalizing (ii) to a similar result for colimit and limit functors involving the following

22.11. Projection functors. Given a small category $\mathbf{D}$, it **terminal projection functor** p_t and its **initial projection functor** p_i will be the functors

$$p_t \colon \mathbf{\Delta} \mathbf{D} \longrightarrow \mathbf{D} \quad \text{and} \quad p_i \colon \mathbf{\Delta}^{\mathrm{op}} \mathbf{D} \longrightarrow \mathbf{D}$$

which send an object $E \colon [n] \to \mathbf{D}$ (22.7) to the objects En and $E0 \in \mathbf{D}$ respectively.

A straightforward calculation then yields that

(i) *for every object $D \in \mathbf{D}$,*

$$(p_t \downarrow D) = \mathbf{\Delta}(\mathbf{D} \downarrow D) \quad \text{and} \quad (D \downarrow p_i) = \mathbf{\Delta}^{\mathrm{op}}(D \downarrow \mathbf{D})$$

and more generally that

(ii) *for every functor $u \colon \mathbf{A} \to \mathbf{B}$ between small categories and object $B \in \mathbf{B}$,*

$$(up_t \downarrow B) = \mathbf{\Delta}(u \downarrow B) \quad \text{and} \quad (B \downarrow up_i) = \mathbf{\Delta}^{\mathrm{op}}(B \downarrow u)$$

and these two observations, together with 19.7(ii) and 22.10(ii) and (iii) then readily imply

22.12. Proposition. *Let $\mathbf{M}$ be a model category. Then, for every small category $\mathbf{D}$, the functor (19.6)*

$$\operatorname{colim}^{p_t} \colon \mathbf{M}^{\mathbf{\Delta} \mathbf{D}} \longrightarrow \mathbf{M}^{\mathbf{D}} \quad (\text{resp. } \lim{}^{p_i} \colon \mathbf{M}^{\mathbf{\Delta}^{\mathrm{op}} \mathbf{D}} \longrightarrow \mathbf{M}^{\mathbf{D}})$$

and more generally, for every functor $u \colon \mathbf{A} \to \mathbf{B}$ between small categories, the functor

$$\operatorname{colim}^{up_t} \colon \mathbf{M}^{\mathbf{\Delta} \mathbf{A}} \longrightarrow \mathbf{M}^{\mathbf{B}} \quad (\text{resp. } \lim{}^{up_i} \colon \mathbf{M}^{\mathbf{\Delta}^{\mathrm{op}} \mathbf{A}} \longrightarrow \mathbf{M}^{\mathbf{B}})$$

sends Reedy weak equivalences between Reedy cofibrant (resp. fibrant) diagrams to objectwise weak equivalences between objectwise cofibrant (resp. fibrant) diagrams.

23. Virtually cofibrant and fibrant diagrams

Given a model category M and a small category D we now introduce the notions of *virtually cofibrant* and *virtually fibrant* D-diagrams in M and show that the full subcategories spanned by them

$$(M^D)_{vc} \subset M^D \quad \text{and} \quad (M^D)_{vf} \subset M^D$$

have the properties mentioned in 20.5.

We start with considering the auxiliary notion of

23.1. Restricted ΔD- and $\Delta^{op} D$-diagrams. Given a model category M and a small category D, we call an object $X \in M^{\Delta D}$ (resp. $M^{\Delta^{op} D}$) (22.10) **restricted** if, for every map $h \in \Delta D$ (resp. $\Delta^{op} D$) for which the associated functor $[n_1] \to [n_2]$ (22.10) sends n_1 to n_2 (resp. 0 to 0), the map $Xh \in M$ is a *weak equivalence* and we denote by

$$(M^{\Delta D})_{res} \subset M^{\Delta D} \quad (\text{resp. } (M^{\Delta^{op} D})_{res} \subset M^{\Delta^{op} D})$$

the full subcategory spanned by these restricted diagrams.

The usefulness of these restricted diagram categories is due to the following proposition which we will prove in 23.5.

23.2. Proposition. *Let M be a model category and D a small category. Then* (22.12)

(i) *the adjunction*

$$\operatorname{colim}^{p_t}\colon (M^{\Delta D})_{res} \longleftrightarrow M^D : p_t^* \quad (\text{resp. } p_i^*\colon M^D \longleftrightarrow (M^{\Delta^{op} D})_{res} : \lim_i^p)$$

is a deformable adjunction (18.2), *which*

(ii) *satisfies the Quillen condition* (18.6) *that, for every Reedy cofibrant (resp. fibrant) object $X \in (M^{\Delta D})_{res}$ (resp. $(M^{\Delta^{op} D})_{res}$) and every object $Y \in M^D$, a map*

$$X \longrightarrow p_t^* Y \in (M^{\Delta D})_{res} \quad (\text{resp. } p_i^* Y \longrightarrow X \in (M^{\Delta^{op} D})_{res})$$

is a Reedy weak equivalence iff its adjunct

$$\operatorname{colim}^{p_t} X \longrightarrow Y \in M^D \quad (\text{resp. } Y \longrightarrow \lim^{p_i} X \in M^D)$$

is an objectwise weak equivalence, and hence (18.6)

(iii) *these functors are inverse homotopical equivalences of homotopical categories* (8.3).

This result in turn suggests the promised notions of

23.3. Virtually cofibrant and fibrant diagrams. Given a model category M and a small category D, we

(i) call an object $Y \in M^D$ **virtually cofibrant** if there exists a Reedy cofibrant object $X \in (M^{\Delta D})_{res}$ such that Y is isomorphic to $\operatorname{colim}^{p_t} X$,

(ii) call an object $Y \in M^D$ **virtually fibrant** if there exists a Reedy fibrant object $X \in (M^{\Delta^{op} D})_{res}$ such that Y is isomorphic to $\lim^{p_i} X$, and

(iii) denote by
$$(M^D)_{vc} \subset M^D \quad \text{and} \quad (M^D)_{vf} \subset M^D$$
the full subcategories of M^D spanned by these virtually cofibrant or fibrant diagrams.

Now we can finally give a

23.4. Proof (of the colimit half) of 20.5. To prove 20.5(i) one notes that in view of 23.2, for every left deformation (r,s) of $M^{\Delta D}$ into $(M^{\Delta D})_c$ (8.3 and 10.3), the pair consisting of the functor
$$\operatorname{colim}^{p_t} rp_t^* \colon M^D \longrightarrow M^D$$
and the natural transformation
$$\operatorname{colim}^{p_t} rp_t^* \longrightarrow 1_{M^D}$$
which associates with every object $Y \in M^D$ the adjunct of the weak equivalence
$$sp_t^* Y \colon rp_t^* Y \longrightarrow p_t^* Y$$
is a left deformation of M^D into $(M^D)_{vc}$ (23.3), while 20.5(ii) readily follows from 22.4 and the observation that the functor $f^{\Delta D} \colon M^{\Delta D} \to N^{\Delta D}$ sends restricted diagrams (23.1) to restricted diagrams.

It thus remains to prove 20.5(iii), i.e. to verify that, for every two Reedy cofibrant diagrams $X_1, X_2 \in (M^{\Delta A})_{res}$ and objectwise weak equivalences
$$f \colon \operatorname{colim}^{p_t} X_1 \longrightarrow \operatorname{colim}^{p_t} X_2 \in M^A$$
the induced map
$$\operatorname{colim}^u f \colon \operatorname{colim}^u \operatorname{colim}^{p_t} X_1 \longrightarrow \operatorname{colim}^u \operatorname{colim}^{p_t} X_2 \in M^B$$
is an objectwise weak equivalence between objectwise cofibrant diagrams. To do this one constructs a commutative diagram in $M^{\Delta A}$ of the form

$$\begin{array}{ccccc}
X_1 & \xrightarrowtail{a}_{\sim} & X_3 & \xleftarrow{c} & X_2 \\
{\scriptstyle \eta_t X_1} \downarrow {\scriptstyle \sim} & & \downarrow {\scriptstyle b} \, {\scriptstyle \sim} & & {\scriptstyle \sim} \downarrow {\scriptstyle \eta_t X_2} \\
p_t^* \operatorname{colim}^{p_t} X_1 & \xrightarrow[\sim]{p_t^* f} & & & p_t^* \operatorname{colim}^{p_t} X_2
\end{array}$$

in which
  (i) the vertical maps are the unit maps, i.e. the adjuncts of the identity maps of $\operatorname{colim}^{p_t} X_1$ and $\operatorname{colim}^{p_t} X_2$, which, in view of 23.2 are weak equivalences
  (ii) maps a and b are (9.1) a factorization of the weak equivalence $(p_t^* f)(\eta_t X_1)$ into a trivial Reedy cofibration a followed by a trivial Reedy fibration b, and
  (iii) the map c is obtained by lifting $\eta_t X_2$ along b.

Adjunction then yields the commutative diagram in M^A

$$\begin{array}{ccccc}
\operatorname{colim}^{p_t} X_1 & \xrightarrow{\operatorname{colim}^{p_t} a} & \operatorname{colim}^{p_t} X_3 & \xleftarrow{\operatorname{colim}^{p_t} c} & \operatorname{colim}^{p_t} X_2 \\
{\scriptstyle 1} \downarrow & & & & \downarrow {\scriptstyle 1} \\
\operatorname{colim}^{p_t} X_1 & & \xrightarrow{f} & & \operatorname{colim}^{p_t} X_2
\end{array}$$

and application of the functor colim^u to this diagram yields a commutative diagram in $\boldsymbol{M}^{\boldsymbol{B}}$ of the form

$$\begin{array}{ccccc}
\operatorname{colim}^u \operatorname{colim}^{p_t} X_1 & \longrightarrow & \operatorname{colim}^u \operatorname{colim}^{p_t} X_3 & \longleftarrow & \operatorname{colim}^u \operatorname{colim}^{p_t} X_2 \\
{\scriptstyle 1}\downarrow & & \searrow & & \downarrow{\scriptstyle 1} \\
\operatorname{colim}^u \operatorname{colim}^{p_t} X_1 & \xrightarrow{\operatorname{colim}^u f} & & & \operatorname{colim}^u \operatorname{colim}^{p_t} X_2
\end{array}$$

in which, in view of (ii), (iii), 19.6(iii) and 22.12, the two top maps are objectwise weak equivalences between objectwise cofibrant objects, which in view of the two out of three property implies that so is the bottom map.

It thus remains to give the promised

23.5. Proof (of the colimit half) of 23.2. For every object $D \in \boldsymbol{D}$, let $p_t^{-1}D \subset \boldsymbol{\Delta D}$ denote the subcategory which has as objects the functors $E\colon [n] \to \boldsymbol{D}$ ($n \geq 0$) such that $En = D$ and as maps $(E_1\colon [n_1] \to \boldsymbol{D}) \to (E_2\colon [n_2] \to \boldsymbol{D})$ the functors $e\colon [n_1] \to [n_2]$ such that $E_1 = E_2 e$ and $en_1 = n_2$. Then one readily verifies that

(i) the category $p_t^{-1}D$ has as an initial object the functor $E_D\colon [0] \to \boldsymbol{D}$ such that $E_D 0 = D$, and is a Reedy category for which

$$\overrightarrow{p_t^{-1}D} = \overrightarrow{\boldsymbol{\Delta D}} \cap p_t^{-1}D \quad \text{and} \quad \overleftarrow{p_t^{-1}D} = \overleftarrow{\boldsymbol{\Delta D}} \cap p_t^{-1}D$$

with the property that $\overleftarrow{p_t^{-1}D}$ is a disjoint union of categories with a terminal object,

and that

(ii) for every Reedy cofibrant diagram $X \in \boldsymbol{M}^{\boldsymbol{\Delta D}}$, the diagram $i^*X \in \boldsymbol{M}^{p_t^{-1}D}$ induced by the inclusion functor $i\colon p_t^{-1}D \to \boldsymbol{\Delta D}$ is also Reedy cofibrant, in view of the fact that, if $d^n\colon [n-1] \to [n]$ denotes the functor such that $d^n i = i$ for $0 \leq i \leq n$, then for every object $E\colon [n] \to \boldsymbol{D} \in p_t^{-1}D$ with $n > 0$, the commutative diagram (22.6)

$$\begin{array}{ccccc}
LX(Ed^n) & \longrightarrow & L(i^*X)E & \longrightarrow & (i^*X)E \\
\downarrow & & \downarrow & & \downarrow{\scriptstyle 1} \\
X(Ed^n) & \longrightarrow & LXE & \longrightarrow & XE
\end{array}$$

has the property that the square on the left is a pushout square and the map $LX(Ed^n) \to X(Ed^n)$ and $XLE \to XE$ are cofibrations and that therefore the map $L(i^*X)E \to (i^*X)E$ is also a cofibration.

One can therefore, for every object $D \in \boldsymbol{D}$, diagram $Y \in \boldsymbol{M}^{\boldsymbol{D}}$, Reedy cofibrant diagram $X \in (\boldsymbol{M}^{\boldsymbol{\Delta D}})_{\text{res}}$ and map $X \to p_t^*Y \in (\boldsymbol{M}^{\boldsymbol{\Delta D}})_{\text{res}}$, consider the induced commutative diagram

$$\begin{array}{ccccccccc}
(\operatorname{colim}^{p_t} X)D & \xrightarrow{a_1} & \operatorname{colim}^{(p_t \downarrow D)} j^* X & \xleftarrow{b_1} & \operatorname{colim}^{p_t^{-1}D} i^* X & \xleftarrow{c_1} & X E_D \\
\downarrow & & \downarrow & & \downarrow & & \downarrow \\
(\operatorname{colim}^{p_t} p_t^* Y)D & \xrightarrow{a_2} & \operatorname{colim}^{(p_t \downarrow D)} j^* p_t^* Y & \xleftarrow{b_2} & \operatorname{colim}^{p_t^{-1}D} i^* p_t^* Y & \xleftarrow{c_2} & p_t^* Y E_D = YD
\end{array}$$

in which

(iii) $j\colon (p_t \downarrow D) = \mathbf{\Delta}(\mathbf{D} \downarrow D) \to \mathbf{\Delta D}$ is the forgetful functor and a_1 and a_2 are the *isomorphism* implied by 19.7(ii),

(iv) b_1 and b_2 are induced by the functor $k\colon p_t^{-1}D \to (p_t \downarrow D)$ which sends an object $E \in p_t^{-1}D$ to the object $(E, 1_{\mathbf{D}}) \in (p_t \downarrow D)$, and are *isomorphisms* because the functor k is right cofinal (i.e. for every object $E \in (p_t \downarrow D)$, the category $(E \downarrow k)$ is non-empty and every two of its objects can be connected by a zigzag of maps), and

(v) c_1 and c_2 are the restrictions to E_D of the respective colimiting cocones and hence ((i) and 22.9) c_1 is a *weak equivalence* while c_2 is actually an *isomorphism*.

A somewhat technical but essentially straightforward calculation then yields that

(vi) the composition $c_2^{-1} b_2^{-1} a_2 \colon (\operatorname{colim}^{p_t} p_t^* Y) D \to Y D$ is the restriction to D of the counit of the adjunction $\operatorname{colim}^{p_t}\colon (\mathbf{M}^{\mathbf{\Delta D}})_{\mathrm{res}} \leftrightarrow \mathbf{M}^{\mathbf{D}} \colon p_t^*$

and that therefore

(vii) the composition of this map with the map $(\operatorname{colim}^{p_t} X) D \to (\operatorname{colim}^{p_t} p_t^* Y) D$ is the restriction to D of the adjunct of the given map $X \to p_t^* Y$

and from this one now readily deduces 23.2.

24. Homotopical comments

We end with noting that the main results of this chapter, i.e. those obtained in §20 and §21, are special cases of, and proven in essentially the same manner, as more general results for cocomplete and complete homotopical categories which we will obtain in chapter VIII.

In more detail:

24.1. Homotopy colimit and limit functors (47.1). Given a homotopical category $\mathbf{X}$ which is cocomplete and complete (19.7) and a map $u\colon \mathbf{A} \to \mathbf{B} \in \mathbf{cat}$,

(i) a **homotopy u-colimit** (resp. **u-limit**) **functor** on $\mathbf{X}$ will be a left (resp. a right) approximation (18.3) of the functor (19.6)

$$\operatorname{colim}^u \colon \mathbf{X}^{\mathbf{A}} \longrightarrow \mathbf{X}^{\mathbf{B}} \qquad (\text{resp. } \lim^u \colon \mathbf{X}^{\mathbf{A}} \longrightarrow \mathbf{X}^{\mathbf{B}}).$$

Thus (13.3)

(ii) *such homotopy u-colimit (resp. u-limit) functors on $\mathbf{X}$, if they exist, are homotopically unique* (13.3)

and essentially the same arguments that (in 20.6) were used to prove 20.2, 20.3 and 20.4 yield that (47.2)

(iii) *the result of 20.2 remains valid if one replaces everywhere the model category $\mathbf{M}$ by a homotopical category $\mathbf{X}$ for which*

(iii)′ *there exist u-colimit (resp. u-limit) functors on $\mathbf{X}$, and*

(iii)″ *the functor colim^u (resp. $\lim^u$)$\colon \mathbf{X}^{\mathbf{A}} \to \mathbf{X}^{\mathbf{B}}$ is left (resp. right) deformable* (18.1),

that (47.4)

(iv) *the result of 20.3 remains valid if one replaces everywhere the model category $\mathbf{M}$ by a homotopical category $\mathbf{X}$ for which*

(iv)′ *there exist u-colimit (resp. u-limit) and v-colimit (resp. v-limit) functors on $\mathbf{X}$, and* (iv)″ *the pair* (colim^u, colim^v) *(resp.* (lim^u, lim^v)*) is left (resp. right) deformable* (18.4)*, which is in particular the case if* $\boldsymbol{X}$ *is saturated* (8.4) *and this pair is locally left (resp. right) deformable* (18.4)

and that (47.6)

(v) *the result of* 20.4 *remains valid if one replaces everywhere the Quillen adjunction* $f\colon \boldsymbol{M} \leftrightarrow \boldsymbol{N} :g$ *by a deformable adjunction* $f\colon \boldsymbol{X} \leftrightarrow \boldsymbol{Y} :g$ (18.2) *for which*

(v)′ *there exist u-colimit (resp. u-limit) functors on both* $\boldsymbol{X}$ *and* $\boldsymbol{Y}$*, and*

(v)″ *the pairs*

$$(\mathrm{colim}^u\colon \boldsymbol{X}^A \to \boldsymbol{X}^B, f^B) \quad \text{and} \quad (f^A, \mathrm{colim}^u\colon \boldsymbol{Y}^A \to \boldsymbol{Y}^B)$$

(resp. $\;(\lim^u\colon \boldsymbol{Y}^A \to \boldsymbol{Y}^B, g^B) \quad \text{and} \quad (g^A, \lim^u\colon \boldsymbol{X}^A \to \boldsymbol{X}^B)\;$ *)*

are left (resp. right) deformable (18.4)*, which is in particular the case if* $\boldsymbol{X}$ *and* $\boldsymbol{Y}$ *are saturated* (8.4) *and these pairs are locally left (resp. right) deformable* (18.4)*.*

It thus remains to discuss

24.2. Homotopical cocompleteness and completeness (49.2). Given a homotopical category $\boldsymbol{X}$ which is cocomplete and complete one

(i) defines **left** and **right cat-systems** and **homotopy colimit** and **limit systems** on $\boldsymbol{X}$ exactly as in 21.1 and 21.3 (i.e. by replacing everywhere in 21.1 and 21.3 the model category $\boldsymbol{M}$ by the homotopical category $\boldsymbol{X}$), and

(ii) calls $\boldsymbol{X}$ **homotopically cocomplete** or **complete** if there exist homotopy colimit or limit systems on $\boldsymbol{X}$.

The arguments used in the proof of 21.4 then yield that (49.3)

(iii) *the result of* 21.4 *remains valid if one replaces everywhere the model category* $\boldsymbol{M}$ *by a homotopical category* $\boldsymbol{X}$ *such that*

(iii)′ $\boldsymbol{X}$ *is cocomplete, complete and saturated* (8.4)*, and*

(iii)″ *there exists, for every object* $\boldsymbol{D} \in \mathbf{cat}$*, a left (resp. a right) deformation retract* $(\boldsymbol{X}^D)_0 \subset \boldsymbol{X}^D$ (8.3) *such that, for every object* $\boldsymbol{E} \in \mathbf{cat}$ *and map* $x\colon \boldsymbol{D} \to \boldsymbol{E} \in \mathbf{cat}$*, the functor* colim^x *(resp.* lim^x*) is homotopical on* $(\boldsymbol{X}^D)_0$*.*

Part II

Homotopical Categories

CHAPTER V

Summary of Part II

25. Introduction

25.1. Motivation. In [**Qui67**] and [**Qui69**] Quillen introduced the notion of a **model category**, i.e. a category with three distinguished classes of maps (*weak equivalences*, *cofibrations* and *fibrations*) satisfying a few simple axioms which enable one to "do homotopy theory". A closer look at this notion however reveals that the weak equivalences already determine the "homotopy theory", while the cofibrations and the fibrations provide additional structure which enables one to "do" homotopy theory, in the sense that, while the homotopy notions involved in doing homotopy theory can be defined in terms of the weak equivalences, the verification of their properties (e.g. their existence) requires the cofibrations and/or the fibrations. Consequently many model category arguments are a mix of arguments which only involve weak equivalences and arguments which also involve cofibrations and/or fibrations and as these two kinds of arguments have different flavors, the resulting mix often looks rather mysterious. In this part II we therefore try to isolate the key model category arguments which involve only weak equivalences by developing a kind of "relative category theory" or theory of **homotopical categories**, which are categories with a single distinguished class of maps called **weak equivalences** which

(i) includes all the *identity maps*, and
(ii) has the **two out of six property** that, for every three maps r, s and t for which the *two* maps sr and ts exist and are weak equivalences, the *four* maps r, s, t and tsr are also weak equivalences,

which property is slightly stronger than Quillen's "two out of three" property.

25.2. Organization of part II. In developing the beginnings of a *relative* or *homotopical category theory* we are guided by the desire to be able to

(i) define *homotopy* or more generally *homotopical colimit* and *limit functors* which, if they exist, are in an appropriate sense *homotopically unique* and the associated notions of *homotopical cocompleteness* and *completeness*, and
(ii) describe useful *sufficient conditions* for the existence of such homotopical colimit and limit functors and for such homotopical cocompleteness and completeness.

In the final chapter VIII we indeed obtain these objectives. However most of the work is done in the two preceding chapters. In the first of these, chapter VI, we introduce *homotopical categories* and *homotopical* (i.e. weak equivalence preserving) *functors* between them, discuss the *homotopy category* of such a homotopical category and explain what exactly we will mean by *homotopical uniqueness*.

In the other chapter, chapter VII, we investigate what we will call *left* and *right deformable functors*, which are a frequently occurring kind of functors between homotopical categories which are *not* necessarily homotopical, but which still have "homotopical meaning", because they have homotopically unique *left* or *right approximations*, i.e. *homotopical* functors which, in a homotopical sense, are closest to them from the left or from the right.

In order to help the reader navigate through all this we devote this first chapter of part II to a preview of some of the key notions and results of chapters VI, VII and VIII.

25.3. Organization of the present chapter. There are five more sections, of which the first three deal with chapter VI and the other two with the other two chapters.

In §26 we explain why we will *not* assume that a category necessarily has *small* hom-sets, introduce *homotopical categories* and *homotopical functors* and discuss a few immediate consequences of their definitions. In §27 we then discuss the *homotopy category* of such a homotopical category and consider in particular the not infrequently occurring case that the homotopical category admits a so-called 3-*arrow calculus* and in §28 we deal with the notion of *homotopical uniqueness*.

In §29 we introduce *deformable functors* and their *approximations* and describe sufficient conditions on a composable pair of deformable functors in order that their composition is also deformable and all compositions of their approximations are approximations of their composition.

And in §30 we then finally discuss *homotopy* and *homotopical colimit functors* and the associated notion of *homotopical cocompleteness*.

26. Homotopical categories

Before introducing homotopical categories we explain what exactly we will mean by

26.1. Categories, locally small categories and small categories. In order to avoid set theoretical difficulties one often defines categories in terms of a *universe*, i.e.

> (i) one assumes that *every set is an element of some universe*, where one defines a **universe** as a set $\mathcal{U}$ of sets (called $\mathcal{U}$-**sets**) satisfying a few simple axioms which imply that $\mathcal{U}$ is closed under the usual operations of set theory and that every $\mathcal{U}$-set is a *subset* of $\mathcal{U}$, but that the set $\mathcal{U}$ itself and many of its subsets are *not* $\mathcal{U}$-sets, and
>
> (ii) one then chooses an *arbitrary but fixed universe* $\mathcal{U}$ and defines a $\mathcal{U}$-**category** as a category of which the hom-sets are $\mathcal{U}$-sets and the set of objects is a *subset* of $\mathcal{U}$ and calls such a $\mathcal{U}$-category **small** if the set of its objects is actually a $\mathcal{U}$-*set*.

It turns out that the notion of a small $\mathcal{U}$-category is indeed a convenient one, in the sense that any "reasonable" operation, when applied to small $\mathcal{U}$-categories yields again a small $\mathcal{U}$-category, but that the notion of a $\mathcal{U}$-category is not, as such operations as the formation of the category of functors between two $\mathcal{U}$-categories or the localization of a $\mathcal{U}$-category with respect to a subcategory is not necessarily again a $\mathcal{U}$-category. Still one cannot avoid the notion of a $\mathcal{U}$-category as many of the categories one is interested in are not small. However a way of getting around this problem is by not insisting on working in only one universe and noting that

(iii) there exists a unique **successor universe** $\mathcal{U}^+$ of our chosen (ii) universe $\mathcal{U}$, i.e. a unique smallest universe $\mathcal{U}^+$ such that the set $\mathcal{U}$ is a $\mathcal{U}^+$-set.

This implies that *every $\mathcal{U}$-category is a small $\mathcal{U}^+$-category* so that any "reasonable" operation when applied to $\mathcal{U}$-categories yields a small $\mathcal{U}^+$-category.

In view of all this **we choose an arbitrary but fixed universe $\mathcal{U}$** and use the term

small category	for *small $\mathcal{U}$-category*, and
locally small category	for *$\mathcal{U}$-category*

(instead of the customary use of the term category for $\mathcal{U}$-category) and reserve, **unless the context clearly indicates otherwise**, the term

category	for *small $\mathcal{U}^+$-category*

Similarly we use the term

small set	for *$\mathcal{U}$-set, and*
set	for *$\mathcal{U}^+$-set.*

26.2. Homotopical categories and homotopical functors. A **homotopical category** will be a category $\boldsymbol{C}$ (26.1) with a distinguished set $\boldsymbol{W}$ (26.1) of maps (called **weak equivalences**) such that

(i) $\boldsymbol{W}$ contains all the *identity maps* of $\boldsymbol{C}$, and
(ii) $\boldsymbol{W}$ has the **two out of six property** that, for every three maps r, s and $t \in \boldsymbol{C}$ for which the *two* compositions sr and ts exist and are in $\boldsymbol{W}$, the *four* maps r, s, t and tsr are also in $\boldsymbol{W}$

which one readily verifies (by assuming that both sr and ts are identity maps or that at least one of r, s and t is an identity map) is equivalent to requiring that

(i)′ $\boldsymbol{W}$ contains all the *identity maps* of $\boldsymbol{C}$,
(ii)′ $\boldsymbol{W}$ has the **weak invertibility property** that every map $s \in \boldsymbol{C}$ for which there exist maps r and $t \in \boldsymbol{C}$ such that the compositions sr and ts exist and are in $\boldsymbol{W}$, is itself in $\boldsymbol{W}$ (which, together with (i)′, implies that all *isomorphisms* of $\boldsymbol{C}$ are in $\boldsymbol{W}$), and
(ii)″ $\boldsymbol{W}$ has the **two out of three property** that, for every two maps f and $g \in \boldsymbol{C}$ for which gf exists and two of f, g and gf are in $\boldsymbol{W}$, so is the third (which implies that $\boldsymbol{W}$ is actually a *subcategory* of $\boldsymbol{C}$).

This definition suggests, given two homotopical categories $\boldsymbol{C}$ and $\boldsymbol{D}$

(iii) calling a functor $f\colon \boldsymbol{C} \to \boldsymbol{D}$ a **homotopical functor** whenever f *preserves weak equivalences*, and
(iv) considering *two* kinds of **homotopical functor** or **diagram categories**, the *first*, denoted by

$$\mathrm{Fun}(\boldsymbol{C}, \boldsymbol{D}) \quad \text{or} \quad \boldsymbol{D}^{\boldsymbol{C}}$$

being the usual functor category (which has as objects the (ordinary) functors $\boldsymbol{C} \to \boldsymbol{D}$ and as maps the natural transformations between them) in which the *weak equivalences* will be the **natural weak equivalences**, i.e. those natural transformations which send the objects of C to weak equivalences in D, while the *other* kind of homotopical functor category will be the full homotopical subcategory of the first spanned by the *homotopical* functors $C \to D$ and will be denoted by

$$\operatorname{Fun}_w(C, D) \quad \text{or} \quad \left(D^C\right)_w .$$

It is also convenient to consider the following notions.

26.3. Homotopical equivalences of homotopical categories. A homotopical functor $f\colon C \to C'$ will be called a **homotopical equivalence of homotopical categories** if there exists a homotopical functor $f'\colon C' \to C$ (called **homotopical inverse** of f) such that the compositions $f'f$ and ff' are **naturally weakly equivalent**, i.e. can be connected by a zigzag of natural weak equivalences, to the identity functors 1_C and $1_{C'}$ respectively.

26.4. Minimal and maximal homotopical categories. Homotopical categories can be seen as a generalization of categories by considering an (ordinary) category as either a *minimal* or a *maximal* homotopical category, i.e. a homotopical category in which either only the isomorphism or all the maps are weak equivalences. The first of these approaches seems to be the more useful one and we therefore

(i) usually consider ordinary categories as *minimal* homotopical categories, and
(ii) when we talk of a **homotopical version** of some categorical notion or result we usually mean (as for instance in §28 below) a notion or result which on *minimal* homotopical categories reduces to the given categorical one.

26.5. The homotopy category of a homotopical category. With a homotopical category C one can, in a functorial manner, associate not only its *underlying category* and its *category of weak equivalences*, but also its **homotopy category** $\operatorname{Ho} C$, i.e. the category obtained from C by "formally inverting" the weak equivalences. In view of the assumptions made in 26.1, such homotopy category always exists (although the homotopy category of a locally small homotopical category need *not* again be locally small).

These homotopy categories and the induced functors between them have the properties that

(i) *if two homotopical functors $f, g\colon C \to D$ between homotopical categories are naturally weakly equivalent, then the induced functors*

$$\operatorname{Ho} f, \operatorname{Ho} g\colon \operatorname{Ho} C \longrightarrow \operatorname{Ho} D$$

are naturally isomorphic, and hence
(ii) *if $f\colon C \to D$ is a homotopical equivalence of homotopical categories, then the induced functor $\operatorname{Ho} f\colon \operatorname{Ho} C \to \operatorname{Ho} D$ is an equivalence of categories.*

27. The hom-sets of the homotopy categories

27.1. An initial description of the hom-sets of the homotopy category. Given a homotopical category C, its *homotopy category* $\operatorname{Ho} C$ (26.5) is obtained from C by "formally inverting" the weak equivalences, i.e. (see 33.8) the category which has the same objects as C and in which, for every two objects

$X, Y \in \boldsymbol{C}$, a map $X \to Y \in \operatorname{Ho}\boldsymbol{C}$ is an appropriately defined equivalence class of **restricted zigzags** in $\boldsymbol{C}$, i.e. sequences of maps in $\boldsymbol{C}$

$$X = \boldsymbol{C}_0 \relbar\joinrel\relbar \boldsymbol{C}_1 \relbar\joinrel\relbar \cdots \relbar\joinrel\relbar \boldsymbol{C}_n = Y \qquad n \geq 0$$

in which some maps are **forward** (i.e. go to to the right), while the others are **backward** (i.e. go to the left) and are *weak equivalences*. It comes with a **localization functor** $\gamma\colon \boldsymbol{C} \to \operatorname{Ho}\boldsymbol{C}$ which is the identity on objects and which sends each map of $\boldsymbol{C}$ to the class of zigzags containing the zigzag which consists of only this map.

This description in which each equivalence class contains zigzags of countably many different **types** (i.e. numbers of forward and backward maps and/or the order in which they appear) can be used to obtain a sometimes more useful

27.2. Colimit description of the hom-sets of the homotopy category. This is a description of each hom-set of $\operatorname{Ho}\boldsymbol{C}$ as a colimit of a diagram of sets, indexed by a **category of types**, in which each set consists of equivalence classes of zigzags in $\boldsymbol{C}$ of only one type.

Occasionally one can do even much better than this and describe the hom-sets themselves as equivalences classes of zigzags of only one type. In particular this happens when the homotopical category $\boldsymbol{C}$ admits a

27.3. 3-arrow calculus. By this we mean that the category of weak equivalences $\boldsymbol{W}$ of $\boldsymbol{C}$ has subcategories $\boldsymbol{U}$ and $\boldsymbol{V}$ (like the categories of the trivial cofibrations and the trivial fibrations in a model category (§11)) such that, in a *functorial* manner,

(i) for every zigzag $A' \xleftarrow{u} A \xrightarrow{f} B$ in $\boldsymbol{C}$ with $u \in \boldsymbol{U}$, there exists a zigzag $A' \xrightarrow{f'} B' \xleftarrow{u'} B$ in $\boldsymbol{C}$ with $u' \in \boldsymbol{U}$ such that $u'f = f'u$ and u' is an isomorphism whenever u is,
(ii) for every zigzag $X \xrightarrow{g} Y \xleftarrow{v} Y'$ in $\boldsymbol{C}$ which $v \in \boldsymbol{V}$, there exists a zigzag $X \xleftarrow{v'} X' \xrightarrow{g'} Y'$ in $\boldsymbol{C}$ with $v' \in \boldsymbol{V}$ such that $gv' = vg'$ and v' is an isomorphism whenever v is, and
(iii) every map $w \in \boldsymbol{W}$ admits a factorization $w = vu$ with $u \in \boldsymbol{U}$ and $v \in \boldsymbol{V}$.

Then we can formulate

27.4. A 3-arrow description of the hom-sets. *If a homotopical category $\boldsymbol{C}$ admits a 3-arrow calculus, then the maps of $\operatorname{Ho}\boldsymbol{C}$ be be described as equivalence classes of restricted zigzags (27.1) in $\boldsymbol{C}$ of the form*

in which two such zigzags are in the same equivalence class iff they are the top row and the bottom row in a commutative diagram in $\boldsymbol{C}$ of the form

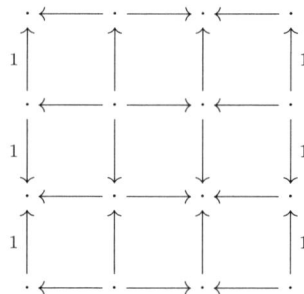

in which the middle rows are also restricted zigzags.

Moreover, using for the first time the full strength of the two out of six property (26.2) and not merely the two out of three property, we deduce from the 3-arrow description the following somewhat surprising

27.5. Saturation result. *Let $\boldsymbol{C}$ be a homotopical category which admits a 3-arrow calculus (27.3). Then $\boldsymbol{C}$ is **saturated** in the sense that a map in $\boldsymbol{C}$ is a weak equivalence iff its image in $\operatorname{Ho}\boldsymbol{C}$ is an isomorphism.*

This last result is not only pleasing, but also quite useful as it allows one, in the many situations (like those which arise from model categories) where one has a 3-arrow calculus, to use the labor saving result mentioned in 29.6, 30.1(v)″ and 30.2(iii) below.

28. Homotopical uniqueness

When working with homotopical categories one often constructs objects in a homotopical category which are *homotopically unique*, in the sense that they satisfy a homotopical version (26.4) of the categorical notion of *uniqueness up to a (unique) isomorphism*, which one encounters when certain objects in a category $\boldsymbol{Y}$ (for instance because they have a common universal property) have the property that, for every two of them Y_1 and Y_2, there is *exactly one* map $Y_1 \to Y_2 \in \boldsymbol{Y}$ and this unique map is moreover an *isomorphism*. We will refer to this categorical notion as *categorical uniqueness* and reformulate it as follows in a fashion which readily points to a corresponding homotopical notion.

28.1. Categorical uniqueness. Given a nonempty set of objects in a category $\boldsymbol{C}$, the objects in this set are **categorically unique** or **canonically isomorphic** if

(i) the *full* subcategory of $\boldsymbol{C}$ spanned by these objects,

or equivalently

(i)′ the **categorically full** subcategory of $\boldsymbol{C}$ spanned by these objects, i.e. the full subcategory of $\boldsymbol{C}$ spanned by these objects and all isomorphic ones,

is **categorically contractible** in the sense that

(ii) this category is a *non-empty groupoid* in which there is exactly *one* isomorphism between any two objects or equivalently

> (ii)′ the unique functor from this category to the category [0] which consists of only a single object 0 and its identity map, is an *equivalence of categories*.

This suggests the following notion of

28.2. Homotopical uniqueness. Given a non-empty set of objects in a homotopical category C, the objects in this set will be called **homotopically unique** or **canonically weakly equivalent** if the **homotopically full** subcategory $G \subset C$ spanned by these objects, i.e. the full subcategory of C spanned by these objects and all weakly equivalent ones is **homotopically contractible** in the sense that the unique functor $G \to [0]$ (28.1) is a *homotopical equivalence of homotopical categories* (26.3).

28.3. Homotopically initial and terminal objects. We also obtain a homotopical version (26.4) of the fact that,

> (i) as *the full subcategory of a category spanned by its initial or terminal objects is empty or categorically contractible* (28.1), one can often verify categorical uniqueness (28.1) by noting that the objects in question are either initial or terminal

and we do this by

> (ii) defining **homotopically initial** or **terminal objects** in a homotopical category with the property that *the full subcategory of a homotopical category spanned by its homotopically initial or terminal objects is a homotopically full* (28.2) *subcategory which is empty or homotopically contractible* (28.2) (the proof of which is the second place where we use the full strength of the two out of six property (26.2)).

29. Deformable functors

In dealing with homotopical categories one often runs into functors between homotopical categories which are *not* necessarily homotopical, but which still have "homotopical meaning" because they are *homotopical* on a so-called *left* (or *right*) *deformation retract* of the domain category and which we will therefore refer to as *left* (or *right*) *deformable functors*.

More precisely

29.1. Left and right deformation retracts. Given a homotopical category X, a **left** (resp. a **right**) **deformation retract** of X will be a *full* subcategory $X_0 \subset X$ for which there exists a **left** (resp. a **right**) **deformation** of X into X_0, i.e. a pair (r, s) consisting of

> (i) a *homotopical* functor $f \colon X \to X$ which sends all of X into X_0, and
> (ii) a *natural weak equivalence*

$$s \colon r \to 1_X \qquad (\text{resp. } s \colon 1_X \to r)$$

(which clearly implies that the inclusion $X_0 \to X$ is a *homotopical equivalence of homotopical categories* (26.3)).

29.2. Left and right deformable functors. A (not necessarily homotopical) functor $f\colon X \to Y$ between homotopical categories will be called **left** (resp. **right**) **deformable** if f is *homotopical* on a *left* (resp. a *right*) *deformation retract* of X (29.1).

The usefulness of this notion is due to the fact that such left and right deformable functors have *homotopically unique* (28.2)

29.3. Left and right approximations. Given a (not necessarily homotopical) functor $f\colon X \to Y$ between homotopical categories, a **left** (resp. a **right**) **approximation** of f will be a pair (k, a) consisting of

  (i) a *homotopical* functor $k\colon X \to Y$, and
  (ii) a natural transformation

$$a\colon k \to f \qquad (\text{resp. } a\colon f \to k)$$

which is a *homotopically terminal* (resp. *initial*) (28.3) object in "the category of homotopical functors $X \to Y$ over (resp. under) f". Thus (28.3(ii))

  (iii) *such a left (resp. right) approximation of f, if it exists, is homotopically unique* (28.2).

Moreover a *sufficient condition* for the existence of such an approximation is that f be *left* (resp. *right*) *deformable*, as we show that

  (iv) *for every left (resp. right) deformation retract $X_0 \subset X$ on which f is homotopical and every left (resp. right) deformation (r, s) of X into X_0* (29.1), *the pair (fr, fs) is a left (resp. a right) approximation of f.*

29.4. Compositions of deformable functors. Given two composable left (or right) deformable functors $f_1\colon X \to Y$ and $f_2\colon Y \to Z$ (29.2), the question arises whether

  (i) their composition $f_2 f_1\colon X \to Z$ is also left (or right) deformable, and
  (ii) the compositions of their left (or right) approximations (29.3) are left (or right) approximations of their composition,

and we show that a *sufficient condition* for (i) and (ii) to happen is that the pair (f_1, f_2) is *left* (or *right*) *deformable* in the following sense.

29.5. Left (or right) deformable composable pairs of functors. Given two composable functors $f_1\colon X \to Y$ and $f_2\colon Y \to Z$ between homotopical categories, the pair (f_1, f_2) will be called **locally left** (resp. **right**) **deformable** if there exist left (resp. right) deformation retracts $X_0 \subset X$ and $Y_0 \subset Y$ (29.1) such that

  (i) f_1 is homotopical on X_0 (and hence f_1 is left (resp. right) deformable),
  (ii) f_2 is homotopical on Y_0 (and hence f_2 is left (resp. right) deformable), and
  (iii) $f_2 f_1$ is also homotopical on X_0 (and hence $f_2 f_1$ is also left (resp. right) deformable),

and the pair (f_2, f_1) will be called **left** (resp. **right**) **deformable** if in addition

  (iv) f_1 sends all of X_0 into Y_0 (which together with (i) and (ii) implies (iii)).

29.6. A special case. Occasionally, to prove the *deformability* of a composable pair of functors, i.e. the existence of deformation retracts satisfying 29.5(i)–(iv), it suffices to verify the *local deformability* of the pair, i.e. the existence of deformation retracts satisfying only 29.5(i)–(iii), because

(i) *given two composable adjunctions*

$$f_1\colon \boldsymbol{X} \longleftrightarrow \boldsymbol{Y} \colon f_1' \qquad \text{and} \qquad f_2\colon \boldsymbol{Y} \longleftrightarrow \boldsymbol{Z} \colon f_2'$$

of functors between homotopical categories such that the pair (f_1, f_2) is locally left deformable (29.5) and the pair (f_2', f_1') is locally right deformable,

(ii) *if the category $\boldsymbol{Z}$ is saturated (27.5) and the pair (f_2', f_1') is right deformable, then the pair (f_1, f_2) is left deformable, and dually*

(iii) *if the category $\boldsymbol{X}$ is saturated and the pair (f_1, f_2) is left deformable, then the pair (f_2', f_1') is right deformable.*

30. Homotopy colimit and limit functors and homotopical ones

As an application of the above results we

(i) generalize the homotopy colimit and limit functors on model categories (see chapter IV) encountered in homotopical algebra, by similarly defining *homotopy colimit* and *limit functors* on a homotopical category $\boldsymbol{X}$ as respectively the left and right approximations (29.1) of arbitrary but fixed colimit and limit functors on $\boldsymbol{X}$, and give sufficient conditions for their existence and composability,

(ii) describe associated notions of *homotopical cocompleteness* and *completeness* which are considerably stronger than the requirement that there exist small homotopy colimit and limit functors and give sufficient conditions for such homotopical cocompleteness and completeness, and

(iii) note that, while the above homotopy colimit and limit functors are only defined when there exist colimit and limit functors, there is a more general notion of what we will call *homotopical colimit* and *limit functors* which is not subject to such a restriction.

In more detail (in the colimit case):

30.1. Homotopy colimit functors. Given a homotopical category $\boldsymbol{X}$ and a category $\boldsymbol{D}$, we

(i) define a **$\boldsymbol{D}$-colimit functor** on $\boldsymbol{X}$ as a left adjoint of the constant diagram functor $c^*\colon \boldsymbol{X} \to \boldsymbol{X}^{\boldsymbol{D}}$ and, if such a $\boldsymbol{D}$-colimit functor exists, denote by $\operatorname{colim}^{\boldsymbol{D}}\colon \boldsymbol{X}^{\boldsymbol{D}} \to \boldsymbol{X}$ an arbitrary but fixed such $\boldsymbol{D}$-colimit functor, and

(i)' define a **homotopy $\boldsymbol{D}$-colimit functor** on $\boldsymbol{X}$ as a left approximation (29.1) of the functor $\operatorname{colim}^{\boldsymbol{D}}$,

and more generally, given a functor $u\colon \boldsymbol{A} \to \boldsymbol{B}$, we

(ii) define a **u-colimit functor** on $\boldsymbol{X}$ as a left adjoint of the induced diagram functor $u^*\colon \boldsymbol{X}^{\boldsymbol{B}} \to \boldsymbol{X}^{\boldsymbol{A}}$ and, if such a u-colimit functor exists, denote by $\operatorname{colim}^u\colon \boldsymbol{X}^{\boldsymbol{A}} \to \boldsymbol{X}^{\boldsymbol{B}}$ an arbitrary but fixed such u-colimit functor, and

(ii)' define a **homotopy u-colimit functor** on $\boldsymbol{X}$ as a left approximation of the functor colim^u.

Then
- (iii) such homotopy u-colimit functors on $\boldsymbol{X}$, if they exist, are *homotopically unique* (28.2), and
- (iv) a sufficient condition for their existence is that there exists a u-colimit functor on $\boldsymbol{X}$ which is *left deformable* (29.2),

and, given two composable functors $u\colon \boldsymbol{A} \to \boldsymbol{B}$ and $v\colon \boldsymbol{B} \to \boldsymbol{D}$

- (v) a sufficient condition in order that every composition of a homotopy u-colimit functor on $\boldsymbol{X}$ with a homotopy v-colimit functor on $\boldsymbol{X}$ is a homotopy vu-colimit functor on $\boldsymbol{X}$

is that

- (v)' the pair $(\operatorname{colim}^u, \operatorname{colim}^v)$ exists and is *left deformable* (29.5)

which in particular is the case if (29.6)

- (v)'' $\boldsymbol{X}$ is *saturated* (27.5) and the pair $(\operatorname{colim}^u, \operatorname{colim}^v)$ exists and is *locally left deformable* (29.5).

30.2. Homotopical cocompleteness. While (categorical) cocompleteness, i.e. the existence of $\boldsymbol{D}$-colimit functors on a homotopical category $\boldsymbol{X}$ for every object $\boldsymbol{D} \in \mathbf{cat}$, where $\mathbf{cat}$ denotes the category of *small* categories (26.1), implies

- (i) the existence, for every map $u \in \mathbf{cat}$, of u-colimit functors on $\boldsymbol{X}$ (29.1(ii)), and
- (i)' the existence, for every composable pair of maps $u, v \in \mathbf{cat}$, u-colimit functor f on $\boldsymbol{X}$, v-colimit functor g on $\boldsymbol{X}$ and vu-colimit functor h on $\boldsymbol{X}$, of a canonical natural *isomorphism* $gf \to h$, and hence
- (i)'' the existence of what we will call a **colimit system** on $\boldsymbol{X}$, i.e. a function F which assigns to every map $u \in \mathbf{cat}$ a u-colimit functor Fu and to every composable pair of maps $u, v \in \mathbf{cat}$, this canonical natural *isomorphism* $(Fv)(Fu) \to F(vu)$ which is *associative* in the obvious sense that every three composable maps $u, v, w \in \mathbf{cat}$ give rise to a commutative diagram of the form

$$\begin{array}{ccc} (Fw)(Fv)(Fu) & \longrightarrow & F(wv)(Fu) \\ \downarrow & & \downarrow \\ (Fw)(Fvu) & \longrightarrow & F(wvu) \end{array},$$

and in particular of the colimit system $\operatorname{colim}^{(\mathbf{cat})}$ which assigns to every map $u \in \mathbf{cat}$ the functor colim^u (30.1)

the existence, for every object $\boldsymbol{D} \in \mathbf{cat}$, of a homotopy $\boldsymbol{D}$-colimit functor on $\boldsymbol{X}$ has in general no such implications.

Therefore we will call a homotopical category $\boldsymbol{X}$ **homotopically cocomplete** if there exists what we will call a **homotopy colimit system** on $\boldsymbol{X}$, which we define as a kind of *left approximation* of the colimit system $\operatorname{colim}^{(\mathbf{cat})}$ (i)'' and which essentially is a function K which assigns to every map $u \in \mathbf{cat}$ a homotopy u-colimit functor on $\boldsymbol{X}$ and to every composable pair of maps $u, v \in \mathbf{cat}$ a natural *weak equivalence* which is associative in the sense of (i)''.

We then show that a *sufficient* condition for such homotopical cocompleteness, i.e. for the *existence* of such a homotopy colimit system on $\boldsymbol{X}$ is that

(ii) X is *cocomplete* and the colimit system $\operatorname{colim}^{(\mathbf{cat})}$ (i)″ is **left deformable** in the sense that there exists a function F_0 which assigns to every object $D \in \mathbf{cat}$, a left deformation retract $F_0 D \in X^D$ (29.1) such that, for every map $u\colon A \to B \in \mathbf{cat}$

(ii)′ the functor $\operatorname{colim}^u\colon X^A \to X^B$ is homotopical on $F_0 A$, and

(ii)″ the functor $\operatorname{colim}^u\colon X^A \to X^B$ sends all of $F_0 A$ into $F_0 B$,

and note that this in particular is the case if

(iii) X is cocomplete and is *saturated* (27.5) and the colimit system $\operatorname{colim}^{(\mathbf{cat})}$ (i)″ is **locally left deformable** in the sense that there exists a function F_0 as in (ii) which satisfies (ii)′ but not necessarily (ii)″

as in that case the saturation ensures the existence of a (possibly different) such function which satisfies both (ii)′ and (ii)″.

It thus remains to discuss

30.3. Homotopical colimit functors. To get around the fact that (30.1), given a functor $u\colon A \to B$, the homotopy u-colimit functors on a category X were only defined when there existed u-colimit functors on X, we note

(i) that the u-colimit functors on X, i.e. the left adjoints of the induced diagram functor $u^*\colon X^B \to X^A$, are also exactly the *Kan extensions* of the identity functor 1_{X^B} along the functor u^*, and

(ii) that there is a notion of **homotopical Kan extensions** of 1_{X^B} along u^* with the property that, if there exist Kan extensions of 1_{X^B} along u^*, then these homotopical Kan extensions of 1_{X^B} along u^* are essentially the same as the left approximations of an arbitrary but fixed such Kan extension.

It follows that

(iii) if there exist u-colimit functors on X, then these homotopical Kan extensions of 1_{X^B} along u^* are essentially the same as the homotopy u-colimit functors

and we therefore define a **homotopical u-colimit functor** on X as a homotopical Kan extension of 1_{X^B} along u^*.

We also note that in a similar fashion the notion of homotopy colimit systems can be generalized to that of **homotopical colimit systems**.

CHAPTER VI

Homotopical Categories and Homotopical Functors

31. Introduction

31.1. Summary. This introductory chapter on homotopical categories and homotopical functors between them consists essentially of three parts.

(i) After attempting to explain why we will *not* assume that a category necessarily has *small* hom-sets, we introduce **homotopical categories** as such categories with a distinguished class of maps (called **weak equivalences**) satisfying two simple axioms, and discuss a few immediate consequences of their definition.

(ii) Next we investigate the **homotopy category** Ho C of such a homotopical category C, i.e. the category obtained from C by "formally inverting" the weak equivalences (which homotopy category, in view of the above (i) mentioned assumption, always "exists"), and in particular

(ii)′ show that if C admits a so-called **3-arrow calculus**, then the maps in Ho C can be described as equivalence classes of zigzags in C of the form

in which the two backward maps are weak equivalences, from which in turn we deduce that

(ii)″ the presence of such a 3-arrow calculus ensures that the homotopical category C has the useful property of being **saturated** in the sense that a map in C is a weak equivalence *iff* its image in Ho C is an isomorphism.

(iii) And finally we end the chapter with a discussion of homotopical analogs of such basic categorical notions as *uniqueness up to a unique isomorphism* (which we will call *categorical uniqueness*), *universal properties*, *initial* and *terminal objects* and *Kan extensions*.

In more detail:

31.2. Categories, locally small categories and small categories. In order to avoid set theoretical difficulties one often defines categories in terms of a *universe*, i.e.

(i) one assumes that *every set is an element of some universe*, where one defines a **universe** as a set $\mathcal{U}$ of sets (called $\mathcal{U}$-**sets**) satisfying a few simple axioms which imply that $\mathcal{U}$ is closed under the usual operations of set theory and that every $\mathcal{U}$-set is a *subset* of $\mathcal{U}$, but that the set $\mathcal{U}$ itself and many of its subsets are *not* $\mathcal{U}$-sets, and

(ii) one then defines a $\mathcal{U}$-**category** as a category of which the hom-sets are $\mathcal{U}$-sets and the set of objects is a *subset* of $\mathcal{U}$ and calls such a $\mathcal{U}$-category **small** if the set of its objects is actually a $\mathcal{U}$-*set*.

It turns out that the notion of a small $\mathcal{U}$-category is indeed a convenient one, in the sense that any "reasonable" operation, when applied to *small* $\mathcal{U}$-*categories* yields again a *small* $\mathcal{U}$-*category*, but that the notion of a $\mathcal{U}$-category is not, as such operations as the formation of the category of functors between two $\mathcal{U}$-categories (33.2) or the localization of a $\mathcal{U}$-category with respect to a subcategory (33.9) do not necessarily yield again a $\mathcal{U}$-category. Still one cannot avoid the notion of a $\mathcal{U}$-category as many of the categories one is interested in are not small. However a way of getting around this problem is by not insisting on working in only one universe and noting that

(iii) for every universe $\mathcal{U}$ there exists a unique **successor universe** $\mathcal{U}^+$, i.e. a unique smallest universe $\mathcal{U}^+$ such that the set $\mathcal{U}$ is a $\mathcal{U}^+$-set.

This implies that *every* $\mathcal{U}$-*category is a small* $\mathcal{U}^+$-*category* so that any "reasonable" operation when applied to $\mathcal{U}$-categories yields a small $\mathcal{U}^+$-category.

In view of all this we choose an **arbitrary but fixed universe** $\mathcal{U}$ and use the term

small category	for small $\mathcal{U}$-category
locally small category	for $\mathcal{U}$-category, and
category	for small $\mathcal{U}^+$-category

(instead of the customary use of the term "category" for $\mathcal{U}$-category) and similarly use the term

small set	for $\mathcal{U}$-set, and
set	for $\mathcal{U}^+$-set

31.3. Homotopical categories. A **homotopical category** will be a category $\boldsymbol{C}$ (31.2) with a distinguished set (31.2) $\boldsymbol{W}$ of maps (called **weak equivalences**) such that

(i) $\boldsymbol{W}$ contains all the *identity maps* of $\boldsymbol{C}$, and
(ii) $\boldsymbol{W}$ has the **two out of six property** that, for every three maps r, s and $t \in \boldsymbol{C}$ for which the *two* compositions sr and ts exist and are in $\boldsymbol{W}$, the *four* maps r, s, t and tsr are also in $\boldsymbol{W}$,

which implies that

(iii) $\boldsymbol{W}$ contains all the isomorphisms of $\boldsymbol{C}$, and
(iv) $\boldsymbol{W}$ has the *two out of three property* (26.2(ii)″) and hence is a *subcategory* of $\boldsymbol{C}$.

Furthermore, given two homotopical categories $\boldsymbol{C}$ and $\boldsymbol{D}$,

(v) a functor $\boldsymbol{C} \to \boldsymbol{D}$ will be called **homotopical** if it *preserves weak equivalences*,
(vi) a natural transformation between two functors $\boldsymbol{C} \to \boldsymbol{D}$ will be called a **natural weak equivalence** if it sends the objects of $\boldsymbol{C}$ to *weak equivalences* in $\boldsymbol{D}$,
(vii) two functors $\boldsymbol{C} \to \boldsymbol{D}$ will be called **naturally weakly equivalent** if they can be connected by a *zigzag* of natural weak equivalences, and

(viii) a homotopical functor $f\colon \boldsymbol{C} \to \boldsymbol{D}$ will be called a **homotopical equivalence of homotopical categories** if there exists a homotopical functor $f'\colon \boldsymbol{D} \to \boldsymbol{C}$ such that the compositions $f'f$ and ff' are naturally weakly equivalent to the identity functors $1_{\boldsymbol{C}}$ and $1_{\boldsymbol{D}}$ respectively.

31.4. The homotopy category. Given a homotopical category $\boldsymbol{C}$ (31.3), its **homotopy category** will be the category (31.2) obtained from $\boldsymbol{C}$ by "formally inverting" the weak equivalences, i.e., the category $\operatorname{Ho} \boldsymbol{C}$ with the same objects as $\boldsymbol{C}$ in which, for every pair of objects $X, Y \in \boldsymbol{C}$, the hom-set (31.2) $\operatorname{Ho} \boldsymbol{C}(X, Y)$ consists of equivalence classes of zigzags in $\boldsymbol{C}$ from X to Y in which the backward maps are weak equivalences, by the weakest equivalence relation which puts two such zigzags in the same class

  (i) when one can be obtained from the other by omitting identity maps,
  (ii) when one can be obtained from the other by replacing adjacent maps which go in the same direction by their composition, and
  (iii) when one can be obtained from the other by omitting two adjacent maps which are the same but go in opposite directions.

From this description of $\operatorname{Ho} \boldsymbol{C}(X, Y)$ in which each equivalence class of zigzag contains zigzags of *countably many types* one can deduce sometimes more useful descriptions of $\operatorname{Ho} \boldsymbol{C}(X, Y)$ as a *colimit* of a diagram of sets in which each set consists of equivalence classes of only *one type*.

31.5. The Grothendieck enrichment. Closely related to this colimit description (31.4) is the **Grothendieck enrichment** of a homotopical category $\boldsymbol{C}$ which is a category enriched over categories which has the same objects as $\boldsymbol{C}$ and in which, for every two objects $X, Y \in \boldsymbol{C}$, the associated hom-category has as its objects all the zigzags in $\boldsymbol{C}$ from X to Y in which the backward maps are weak equivalences and of which two such objects can be connected by a zigzag of maps iff they represent the same element in $\operatorname{Ho} \boldsymbol{C}(X, Y)$.

Moreover this Grothendieck enrichment is closely relate to the (hammock) simplicial localization of [**DK80a**].

31.6. 3-arrow calculi. The above (31.4) colimit description of the hom-sets of the homotopy category of a homotopical category $\boldsymbol{C}$ can be considerably simplified if $\boldsymbol{C}$ admits a **3-arrow calculus**, i.e. if the category of weak equivalences $\boldsymbol{W}$ of $\boldsymbol{C}$ has subcategories $\boldsymbol{U}$ and $\boldsymbol{V}$ (like the categories of the trivial cofibrations and the trivial fibrations in a model category (11.4)) such that, in a *functorial* manner,

  (i) for every zigzag $A' \xleftarrow{u} A \xrightarrow{f} B$ in $\boldsymbol{C}$ with $u \in \boldsymbol{U}$, there exists a zigzag $A' \xrightarrow{f'} B' \xleftarrow{u'} B$ in $\boldsymbol{C}$ with $u' \in \boldsymbol{U}$ such that $u'f = f'u$ and in which u' is an isomorphism whenever u is,
  (ii) for every zigzag $X \xrightarrow{g} Y \xleftarrow{v} Y'$ in $\boldsymbol{C}$ which $v \in \boldsymbol{V}$, there exists a zigzag $X \xleftarrow{v'} X' \xrightarrow{g'} Y'$ in $\boldsymbol{C}$ with $v' \in \boldsymbol{V}$ such that $gv' = vg'$ and in which v' is an isomorphism whenever v is, and
  (iii) every map $w \in \boldsymbol{W}$ admits a factorization $w = vu$ with $u \in \boldsymbol{U}$ and $v \in \boldsymbol{V}$.

Such a homotopical category $\boldsymbol{C}$ which admits a 3-arrow calculus then has the property that, for every pair of objects $X, Y \in \boldsymbol{C}$, the hom-set $\operatorname{Ho} \boldsymbol{C}(X, Y)$ can be

described as the set of the equivalence classes of zigzags in C of only the form
$$X \longleftarrow \cdot \longrightarrow \cdot \longleftarrow Y$$
in which the backward maps are weak equivalences, where two such zigzags are in the same class iff they are the top row and the bottom row in a commutative diagram in C of the form

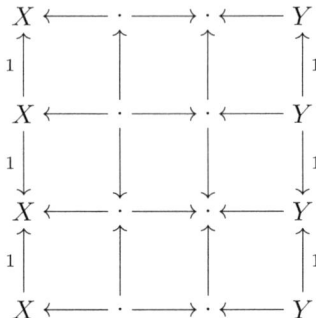

in which the backward maps and hence also the vertical maps are weak equivalences, from which in turn we deduce the rather useful

31.7. Saturation result. Using the above 3-arrow description of the hom-sets we show that *if C is a homotopical category which admits a 3-arrow calculus (31.6), then C is **saturated** in the sense that a map in C is a weak equivalence iff its image in $\operatorname{Ho} C$ is an isomorphism.*

31.8. Categorical and homotopical uniqueness. We reformulate the usual notion of *uniqueness up to a unique isomorphism* of certain objects in a category Y, i.e. the fact that for any two of these objects Y_1 and Y_2 there is *exactly one* map $Y_1 \to Y_2 \in Y$ and that this unique map is an *isomorphism*, by saying that, given a category Y and a non-empty set I (31.2), certain objects Y_i ($i \in I$) are **categorically unique** or **canonically isomorphic** if the *full* subcategory of Y spanned by these objects (or equivalently the **categorically full** subcategory spanned by these objects, i.e. the full subcategory spanned by these objects and all isomorphic ones) is **categorically contractible** in the sense that it is a *non-empty groupoid* in which there is *exactly one* isomorphism between any two objects, or equivalently, that the unique functor from this category to the category [0] (where [0] denotes the category which consists of only one object 0 and its identity map) is an *equivalence of categories*. If Y is a homotopical category we therefore say similarly that objects Y_i ($i \in I$) are **homotopically unique** or **canonically weakly equivalent** if the **homotopically full** subcategory spanned by these objects, i.e. the full subcategory spanned by these objects and all weakly equivalent ones, is **homotopically contractible** in the sense that the unique functor $G \to [0]$ is a *homotopical equivalence of homotopical categories* (31.3).

As

(∗) *the full subcategory of a category spanned by its initial or terminal objects is either empty or categorically contractible* (31.8),

one can often verify the categorical uniqueness (31.8) of a non-empty set of objects in a category Y by noting that these objects are *initial* or *terminal* objects of

Y and similarly one can often verify the homotopical uniqueness (31.8) of a non-empty set of objects in a homotopical category Y by noting that these objects are *homotopically initial* or *homotopically terminal* objects of Y in the following sense.

31.9. Homotopically initial and terminal objects. If, given a category Y and an object $Y \in Y$, one denotes by $1_Y \colon Y \to Y$ the *identity* functor of Y and by $\mathrm{cst}_Y \colon Y \to Y$ the *constant* functor which sends all maps of Y to the identity map of the object $Y \in Y$, then (see 38.1) *an object $Y \in Y$ is initial (resp. terminal) iff there exists a natural transformation*

$$f \colon \mathrm{cst}_Y \to 1_Y \qquad (\text{resp. } f \colon 1_Y \to \mathrm{cst}_Y)$$

such that the map $fY \colon Y \to Y \in Y$ *is an isomorphism.*

We therefore call an object Y of a homotopical category Y **homotopically initial** (or **terminal**) if there exist homotopical functors $F_0, F_1 \colon Y \to Y$ and a natural transformation

$$f \colon F_0 \to F_1 \qquad (\text{resp. } f \colon F_1 \to F_0)$$

such that the map $fY \in Y$ is a *weak equivalence* and the functors F_0 and F_1 are *naturally weakly equivalent* (31.3) to cst_Y and 1_Y respectively, and note that this definition indeed implies that

> (∗) *the full subcategory of Y spanned by the homotopically initial (resp. terminal) objects is a homotopically full subcategory* (31.8) *which is either empty or homotopically contractible* (31.8).

31.10. Homotopical Kan extensions. Just as one defines (ordinary) Kan extensions as terminal or initial objects in an appropriate category, one can define **homotopical Kan extensions** as homotopically terminal or initial objects in an appropriate homotopical category. Except for a brief mention in 39.3 and 41.1, these Kan extensions will only be used in the second half of chapter VIII.

We end with

31.11. Organization of the chapter. After some categorical preliminaries (in §32) we introduce *homotopical categories* and define their *homotopy categories* (in §33). Next (in §34) we give a more explicit description of this homotopy category and investigate (in §36) the consequences of the presence of a *3-arrow calculus*. And finally we discuss *homotopical uniqueness* and *homotopically universal properties* (in §37) and the related notions of *homotopically terminal* and *initial objects* and *homotopical Kan extensions* (in §38).

32. Universes and categories

The aim of this section is to explain what exactly we mean by the terms **category**, **locally small category**, and **small category**.

In order to avoid the usual set theoretical problems we will work in the setting of

32.1. Universes. ([**AGV72**], [**Mac71**], [**Sch72**]). When one works in the setting of universes one makes the basic assumption that

(∗) **every set is an element of some universe**

where a universe is, roughly speaking, a set of "sufficiently small sets" satisfying a few simple axioms which ensure that the usual operations of set theory, when applied to these sufficiently small sets, produce again such sets. More precisely, a **universe** is a set $\mathcal{U}$ of sets (called $\mathcal{U}$-**sets**) such that

(i) if $x \in \mathcal{U}$ and $y \in x$, then $y \in \mathcal{U}$,
(ii) if $x \in \mathcal{U}$ and $y \in \mathcal{U}$, then $\{x,y\} \in \mathcal{U}$, where $\{x,y\}$ denotes the 2-element set with x and y as its elements,
(iii) if $a \in \mathcal{U}$ and, for every $i \in a$, $x_i \in \mathcal{U}$, then the union $\bigcup_{i \in a} x_i \in \mathcal{U}$,
(iv) if $x \in \mathcal{U}$, then $\mathrm{P}(x) \in \mathcal{U}$, where $\mathrm{P}(x)$ denotes the power set of x, i.e. the set of the subsets of x, and
(v) the set of the finite ordinals is an element of $\mathcal{U}$,

from which one can deduce such other closure properties as

(vi) if $x \in \mathcal{U}$ and $y \subseteq x$, then $y \in \mathcal{U}$,
(vii) if $a \in \mathcal{U}$ and, for every $i \in a$, $x_i \in \mathcal{U}$, then the product $\prod_{a \in a} x_i \in \mathcal{U}$,
(viii) if $x \in \mathcal{U}$ and $y \in \mathcal{U}$, then $x^y \in \mathcal{U}$, where x^y denotes the set of the functions $y \to x$,
(ix) the intersection of any set of universes is again a universe, and
(x) if $x \in \mathcal{U}$, then $x \subseteq \mathcal{U}$.

However the opposite of (x) does not hold, i.e. not every subset of $\mathcal{U}$ is an element of $\mathcal{U}$. For instance, $\mathcal{U}$ itself cannot be an element of $\mathcal{U}$ as this would imply (iv) that $\mathrm{P}(\mathcal{U})$ would also be an element of $\mathcal{U}$ and hence (x) a subset of $\mathcal{U}$, which is impossible as $\mathrm{P}(\mathcal{U})$ has greater cardinality than $\mathcal{U}$.

A consequence of (∗) is not only the existence of universes, but also the fact that

(xi) for every universe $\mathcal{U}$, there exists a **higher universe**, i.e. a universe $\mathcal{V}$ such that $\mathcal{U} \in \mathcal{V}$, and in fact (ix) a *unique* smallest such higher universe which will be called the **successor universe** of $\mathcal{U}$ and denoted by $\mathcal{U}^+$.

In such a setting of universes one then defines as follows

32.2. $\mathcal{U}$-**categories and small** $\mathcal{U}$-**categories.** Given a universe $\mathcal{U}$ (32.1), a $\mathcal{U}$-**category** will be a category C such that

(i) for every pair of object $X, Y \in C$, the **hom-set** $C(X,Y)$, i.e. the set of the maps $X \to Y \in C$, is an *element* of $\mathcal{U}$, and
(ii) the set of the objects of C is a *subset* of $\mathcal{U}$,

and a **small** $\mathcal{U}$-**category** will be a $\mathcal{U}$-category such that

(ii)′ the set of the objects of C is actually an *element* of $\mathcal{U}$, which implies (32.1(iii)) that so is the set of *all* its maps.

These definitions readily imply that

(iii) *the small $\mathcal{U}$-categories and the functors between them form a $\mathcal{U}$-category.*

However this conclusion does not hold for the (not necessarily small) $\mathcal{U}$-categories as the set of all $\mathcal{U}$-categories has greater cardinality than $\mathcal{U}$. But

(iv) *every $\mathcal{U}$-category is a small $\mathcal{U}^+$-category,* where $\mathcal{U}^+$ denotes the successor universe of $\mathcal{U}$ (32.1). Thus $\mathcal{U}$-categories and the functors between them form a $\mathcal{U}^+$-category and a rather straightforward calculation yields that in fact

(v) *the $\mathcal{U}$-categories and the functors between them form a small $\mathcal{U}^+$-category.*

Now we can explain what we will mean by

32.3. Categories, locally small categories and small categories. We choose an **arbitrary but fixed universe** $\mathcal{U}$ and use the term

small category	for *small $\mathcal{U}$-category*, and
locally small category	for *$\mathcal{U}$-category*

instead of the customary use of the term "category" for $\mathcal{U}$-categories, and **unless the context clearly indicates otherwise** we reserve the term

category	for *small $\mathcal{U}^+$-category*.

Moreover we use the term

small set	for *$\mathcal{U}$-set*, and
set	for *$\mathcal{U}^+$-set*.

In a similar fashion we denote by

cat	the *$\mathcal{U}$-category of the small $\mathcal{U}$-categories,* and
CAT	the *$\mathcal{U}^+$-category of the small $\mathcal{U}^+$-categories.*

Our justification for the above terminology is the following:

The notion of a small $\mathcal{U}$-category is a quite convenient one as any "reasonable" operation, when applied to small $\mathcal{U}$-categories, yields again a small $\mathcal{U}$-category. For instance, in view of the arguments given in 33.2 and 33.8 below (with $\mathcal{U}$ instead of $\mathcal{U}^+$),

(i) the category of functors between two small $\mathcal{U}$-categories (33.2), and

(ii) the localization of a small $\mathcal{U}$-category with respect to a subcategory (33.9)

are again small $\mathcal{U}$-categories. However many interesting $\mathcal{U}$-categories, and in particular the model categories of part I to which we want to apply the results of part II, are not small and, while many operations, when applied to $\mathcal{U}$-categories yield again a $\mathcal{U}$-category,

(i)′ the colimit or limit of a diagram of $\mathcal{U}$-categories indexed by a $\mathcal{U}$-category,

(ii)′ the category of functors between two $\mathcal{U}$-categories, and

(iii)′ the localization of a $\mathcal{U}$-category with respect to a subcategory

are (in view of 32.2 (iv)) always a small $\mathcal{U}^+$-category, but need not be a $\mathcal{U}$-category. As a result, in dealing with $\mathcal{U}$-categories it is difficult to avoid the use of small $\mathcal{U}^+$-categories and one gets a cleaner exposition if one works most of the time with small $\mathcal{U}^+$-categories and, only in the few instances when this is really necessary, notes sufficient conditions in order that some operation on $\mathcal{U}$-categories yields a small $\mathcal{U}^+$-category which is actually a $\mathcal{U}$-category.

33. Homotopical categories

We now introduce homotopical categories and homotopical functors between them and discuss several functorial ways in which one can associate with each homotopical category an (ordinary) category and conversely with each category a homotopical category.

33.1. Homotopical categories and homotopical functors. A **homotopical category** will be a category C (32.3) with a distinguished set W (32.3) of maps (called **weak equivalences**) such that

(i) W contains all the *identity maps* of C, and
(ii) W has the **two out of six property** that, for every three maps r, s and $t \in C$ for which the *two* compositions sr and ts exist and are in W, the four maps r, s, t and tsr are also in W

which one readily verifies (by assuming that both sr and ts are identity maps or that at least one of r, s and t is an identity map) is equivalent to requiring that

(i) W contains all the identity maps of C,
(ii)′ W has the **weak invertibility property** that every map $s \in C$ for which there exist maps r and $t \in C$ such that the compositions sr and ts exist and are in W, is itself in W (which together with (iii) implies that all *isomorphisms* of C are in W), and
(ii)″ W has the **two out of three property** that, for every two maps f and $g \in C$ for which gf exists and two of f, g and gf are in W, the third is also in W (which implies that W is a *subcategory* of C).

Such a homotopical category will be called **small** or **locally small** whenever the category C is (32.3).

Two objects of C will be called **weakly equivalent** if they can be connected by a *zigzag* of weak equivalences. The category C will be referred to as the **underlying category** and its subcategory W as the **homotopical structure** or **category of weak equivalences** and a **homotopical subcategory** of C will be a subcategory $C_0 \subset C$ with $C_0 \cap W$ as its homotopical structure.

Furthermore, given two homotopical categories C and C'

(iii) a functor $C \to C'$ will be called **homotopical** if it **preserves** weak equivalences, i.e. sends the weak equivalences in C to weak equivalences in C',
(iv) a natural transformation $f \to g$ between two functors $f, g \colon C \to C'$ will be called a **natural weak equivalence** if it sends the objects of C to weak equivalences in C', and
(v) a homotopical functor $f \colon C \to C'$ will be called a **homotopical equivalence of homotopical categories** if there exists a homotopical functor $f' \colon C' \to C$ (called a **homotopical inverse** of f) such that the compositions $f'f$ and ff' are **naturally weakly equivalent**, i.e. can be connected by a *zigzag* of natural weak equivalences, to the identity functors 1_C and $1_{C'}$ respectively.

We also note that functors between homotopical categories and the natural transformations between them give rise to two kinds of

33.2. Homotopical diagram or functor categories. Given two categories C and D (32.3), a C-diagram in D is just a functor $C \to D$ and one sometimes refers to C as the **indexing category** of the diagram. Such diagrams and the natural transformations between them form a category in the sense of 32.3 (as these natural transformations can be considered as functions from the $\mathcal{U}^+$-set of the objects of C to the $\mathcal{U}^+$-set of the maps of D and, in view of 32.1(viii), the set of all such functions is also a $\mathcal{U}^+$-set). This category is called the **functor category** or, especially when the category C plays a rather different role than the category D, **diagram category** and will be denoted by

$$\operatorname{Fun}(C, D) \quad \text{or} \quad D^C .$$

If C and D are *homotopical* categories, then one can similarly form *two* kinds of **homotopical diagram** or **functor categories**. The first of these is the **homotopical category of** (ordinary) **functors** $C \to D$ which has

(i) as objects the (ordinary) functors $C \to D$,
(ii) as maps the natural transformations between them, and
(iii) as weak equivalences the natural weak equivalences,

for which we will use the same notation

$$\operatorname{Fun}(C, D) \quad \text{or} \quad D^C$$

as for the (ordinary) functor category, while the second will be the **homotopical category of homotopical functors** $C \to D$, i.e. its full homotopical subcategory

(33.1)
$$\operatorname{Fun}_{\mathrm{w}}(C, D) \quad \text{or} \quad \left(D^C\right)_{\mathrm{w}}$$

spanned by the *homotopical* functors $C \to D$.

Next we consider

33.3. Adjunctions connecting the category of homotopical categories and homotopical functors with the category of (ordinary) categories. The homotopical categories and the homotopical functors between them (33.1) form a $\mathcal{U}^+$-category (32.1 and 32.2) which we will denote by $\mathbf{CAT}_{\mathrm{w}}$, and which is related to the $\mathcal{U}^+$-category $\mathbf{CAT}$ of (ordinary) categories and functors (32.3) by a sequence of five functors (two in one direction and three in the other) which are connected by four adjunctions as indicated in the diagram

$$\mathrm{Ho} \longleftrightarrow \min \longleftrightarrow j \longleftrightarrow \max \longleftrightarrow j_{\mathrm{w}} ,$$

and we devote the remainder of this section to a brief discussion of these five functors.

We start with

33.4. The forgetful functors $j, j_{\mathrm{w}} \colon \mathbf{CAT}_{\mathrm{w}} \to \mathbf{CAT}$. These are the functors which send a homotopical category respectively to its underlying category and its category of weak equivalences ((33.1)).

Closely related to both these functors is

33.5. The maximal structure functor $\max \colon \mathbf{CAT} \to \mathbf{CAT}_{\mathrm{w}}$. This is the functor which sends a category C to the **maximal homotopical category** with C as its underlying category as well as its category of weak equivalences (33.1). It obviously is (33.4)

(i) *a left adjoint of the forgetful functor* $j_{\mathrm{w}} \colon \mathbf{CAT}_{\mathrm{w}} \to \mathbf{CAT}$, *and*
(ii) *a right adjoint of the forgetful functor* $j \colon \mathbf{CAT}_{\mathrm{w}} \to \mathbf{CAT}$.

Of course there is also the other extreme of

33.6. The minimal structure functor min: **CAT** → **CAT**$_w$. This is the functor which sends a category C to the **minimal homotopical category** which has C as its underlying category, but in which the weak equivalences are only the isomorphisms of C. It clearly is

(∗) *a left adjoint of the forgetful functor* j: **CAT**$_w$ → **CAT** (33.4).

These last two results (33.5 and 33.6) suggest two ways of

33.7. Embedding CAT in CAT$_w$. Homotopical categories can be seen as a generalization of categories by considering an (ordinary) category as either a *minimal* or a *maximal* homotopical category, i.e. (33.6 and 33.5) a homotopical category in which either only the isomorphisms or all the maps are weak equivalences. The first of these approaches seems to be the more useful one and we therefore usually will

(i) consider **CAT** as a full subcategory of **CAT**$_w$ by considering each (ordinary) category as a *minimal* homotopical one, and
(ii) mean, when we talk of a **homotopical version** of some categorical notion or result, a notion or result which (for instance as in §37 and §38 below) on minimal homotopical categories reduces to the given categorical one.

It thus remains to discuss the functor Ho: **CAT**$_w$ → **CAT** which sends homotopical categories to their

33.8. Homotopy categories. Given a homotopical category C, its **homotopy category** will be the category HoC obtained from C by "formally inverting" the weak equivalences, i.e. the category defined as follows.

Given two objects $X, Y \in C$ and an integer $n \geq 0$, a **zigzag** in C from X to Y of **length** n will be a sequence

$$X = C_0 \xrightarrow{f_1} C_1 \underline{\qquad} \cdots \underline{\xleftarrow{f_n}} C_n = Y$$

of maps in C, each of which is either **forward** (i.e. points to the right) or **backward** (i.e. points to the left) and such a zigzag will be called **restricted** if all the backward maps are *weak equivalences*. The **homotopy category** of C then will be the category HoC which has the same objects as C, in which, for every two objects $X, Y \in C$, the hom-set Ho$C(X, Y)$ is the set of the equivalence classes of the *restricted* zigzags in C from X to Y, where two such zigzags are in the same class iff one can be transformed into the other by a finite sequence of operations of the following three types and their inverses:

(i) omitting an identity map,
(ii) replacing two adjacent maps which go in the same direction by their composition, and
(iii) omitting two adjacent maps when they are the same, but go in opposite directions

and in which the compositions are induced by the compositions of the zigzags involved.

It thus remains to verify that

(iv) *the homotopy category* $\operatorname{Ho} C$ *so defined is indeed a category, i.e.* (32.3) *a small* $\mathcal{U}^+$*-category*

but this follows readily from the observation that the set of the maps of $\operatorname{Ho} C$ is a quotient of a subset of the finite ordered subsets of the set of the maps of C and that, as (32.2(ii)′) the set of the maps of C is a $\mathcal{U}^+$-set, so is, in view of 32.1(viii), the set of its finite ordered subsets.

We also note that the same argument with $\mathcal{U}$ instead of $\mathcal{U}^+$ yields that

(v) *the homotopy category of a small homotopical category* (32.3) *is also small.*

However the homotopy category of a locally small homotopical category (32.3) need *not* be locally small.

33.9. The localization functor. Given a homotopical category C with category of weak equivalences W, its homotopy category $\operatorname{Ho} C$ (33.8) comes with a functor $\gamma\colon C \to \operatorname{Ho} C$ which is the identity on the objects and which sends a map $f\colon X \to Y \in C$ to the map $\gamma f\colon X \to Y \in \operatorname{Ho} C$ which contains the restricted zigzag in C from X to Y which consists of only the map f. We will refer to it as the **localization functor** of C as the definitions involved imply that

(i) *the pair* $(\operatorname{Ho} C, \gamma)$ *is a **localization** of* C *with respect to* W *in the sense that this pair has the universal property that, for every category* B *and functor* $b\colon C \to B$ *which sends the weak equivalences in* C *to isomorphisms in* B*, there is a unique functor* $b_\gamma\colon \operatorname{Ho} C \to B$ *such that* $b_\gamma \gamma = b$.

This universal property implies that

(ii) *given two homotopical categories* C *and* D *with localization functors* $\gamma\colon C \to \operatorname{Ho} C$ *and* $\gamma'\colon D \to \operatorname{Ho} D$ *there is, for every homotopical functor* $f\colon C \to D$ *a unique functor* $\operatorname{Ho} f\colon \operatorname{Ho} C \to \operatorname{Ho} D$ *such that* $(\operatorname{Ho} f)\gamma = \gamma' f$ *and, for every natural transformation (resp. natural weak equivalence)* $d\colon f_1 \to f_2$ *between two homotopical functors* $f_1, f_2\colon C \to D$*, a unique natural transformation (resp. natural isomorphism)* $\operatorname{Ho} d\colon \operatorname{Ho} f_1 \to \operatorname{Ho} f_2$ *such that* $(\operatorname{Ho} d)\gamma = \gamma' d$,

(iii) *the function* Ho *so defined is a functor* $\operatorname{Ho}\colon \mathbf{CAT}_\mathrm{w} \to \mathbf{CAT}$ *which is left adjoint to the functor* $\min\colon \mathbf{CAT} \to \mathbf{CAT}_\mathrm{w}$ (33.6), *and*

(iv) *this functor* Ho *sends naturally weakly equivalent homotopical functors* $C \to D$ (33.1(vii)) *to naturally isomorphic functors* $\operatorname{Ho} C \to \operatorname{Ho} D$ *and homotopical equivalences of homotopical categories* $C \to D$ (33.1(vii)) *to equivalences of categories* $\operatorname{Ho} C \to \operatorname{Ho} D$.

Closely related to the localization functor is the rather useful (see 42.5, 45.2, 47.4, 47.6 and 49.3 below) notion of a **saturated** homotopical category, i.e. a homotopical category C with localization functor $\gamma\colon C \to \operatorname{Ho} C$ with the property that a map $c \in C$ is a weak equivalence *iff* the map $\gamma c \in \operatorname{Ho} C$ is an isomorphism. This notion is *hereditary* in the sense that

(v) *if* C *is a saturated homotopical category, then so are, for every homotopical category* D*, the homotopical functor categories* (33.2)

$$C^D \quad \text{and} \quad (C^D)_\mathrm{w}$$

This follows readily from the observation that every object $D \in \boldsymbol{D}$ gives rise to a commutative diagram of the form

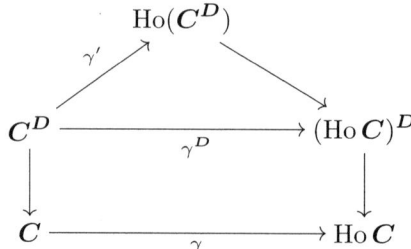

in which the vertical maps are "restrictions to $\boldsymbol{D}$", γ and γ' are the localization functors of $\boldsymbol{C}$ and $\boldsymbol{C}^{\boldsymbol{D}}$ and the map in the upper right hand corner is the unique map (see (i)) such that the triangle commutes.

We end this discussion of homotopy categories with an often convenient

33.10. Alternate description of the hom-sets of the homotopy category. *Given a homotopical category $\boldsymbol{C}$ and a pair of objects $X, Y \in \boldsymbol{C}$, the hom-set $\mathrm{Ho}\,\boldsymbol{C}(X,Y)$ (33.8) is the set of the equivalence classes of the restricted zigzags in $\boldsymbol{C}$ from X to Y, where two such zigzags are in the same class if one can be transformed into the other by a finite sequence of operations of the following two types and their inverses:*

(i) *omitting an identity map, and*
(ii) *in a commutative diagram in $\boldsymbol{C}$ of the form*

$$\begin{array}{ccccccc} X & \xrightarrow{f_1} & \cdot & \cdots & \cdot & \xrightarrow{f_n} & Y \\ {\scriptstyle 1=a_0}\downarrow & & {\scriptstyle a_1}\downarrow & & {\scriptstyle a_{n-1}}\downarrow & & \downarrow{\scriptstyle a_n=1} \\ X & \xrightarrow{g_1} & \cdot & \cdots & \cdot & \xrightarrow{g_n} & Y \end{array} \qquad n \geq 0$$

in which, for every integer i with $1 \leq i \leq n$, the maps f_i and g_i go in the same direction, replacing the top row by the bottom row.

Proof. As every weak equivalence $c\colon C_0 \to C_1$ and every composable pair of maps $c_1\colon C_1 \to C_2$ and $c_2\colon C_2 \to C_3 \in \boldsymbol{C}$ give rise to commutative diagrams

$$\begin{array}{ccccc} C_0 \xrightarrow{c} C_1 \xleftarrow{c} C_0 & \quad & C_1 \xleftarrow{c} C_0 \xrightarrow{c} C_1 & \quad & C_1 \xrightarrow{c_1} C_2 \xrightarrow{c_2} C_3 \\ {\scriptstyle 1}\uparrow \quad {\scriptstyle c}\uparrow \quad \uparrow{\scriptstyle 1} & , & {\scriptstyle 1}\downarrow \quad {\scriptstyle c}\downarrow \quad \downarrow{\scriptstyle 1} & \text{and} & {\scriptstyle 1}\downarrow \quad {\scriptstyle c_2}\downarrow \quad \downarrow{\scriptstyle 1} \\ C_0 \xrightarrow{1} C_0 \xleftarrow{1} C_0 & & C_1 \xleftarrow{1} C_1 \xrightarrow{1} C_1 & & C_1 \xrightarrow{} C_3 \xrightarrow{1} C_3 \end{array}$$

every two restricted zigzags in $\boldsymbol{C}$ which are in the same class in the sense of 33.8 are also in the same class in the sense of 33.10.

To show the converse one has to prove that, for every commutative diagram as in 33.10(ii) and every integer i with $1 \leq i \leq n$, the restricted zigzag

$$X \xrightarrow{f_1} \cdot \cdots \cdot \xrightarrow{f_{i-1}} \cdot \xrightarrow{f_i} \cdot \xrightarrow{a_i} \cdot \xrightarrow{g_{i+1}} \cdot \cdots \cdot \xrightarrow{g_n} Y$$

and

$$X \xrightarrow{f_1} \cdot \cdots \cdot \xrightarrow{f_{i-1}} \cdot \xrightarrow{a_{i-1}} \cdot \xrightarrow{g_i} \cdot \xrightarrow{g_{i+1}} \cdot \cdots \cdot \xrightarrow{g_n} Y$$

are in the same class in the sense of 33.8 and one does this by noting that, in the case that f_i and g_i are *forward* maps, this follows from 33.8(ii) and the fact that $a_i f_i = g_i a_{i-1}$, and that, in the case that f_i and g_i are *backward* maps, this follows from the observation that, in view of 33.8(ii), these two zigzags are respectively in the same class, in the sense of 33.8, as the zigzags

$$X \cdots \cdot \xrightarrow{f_{i-1}} \cdot \xleftarrow{f_i} \cdot \xrightarrow{f_i} \cdot \xrightarrow{a_{i-1}} \cdot \xrightarrow{g_i} \cdot \xrightarrow{g_{i+1}} \cdot \cdots Y$$

and

$$X \cdots \cdot \xrightarrow{f_{i-1}} \cdot \xrightarrow{f_i} \cdot \xrightarrow{a_i} \cdot \xrightarrow{g_i} \cdot \xleftarrow{g_i} \cdot \xrightarrow{g_{i+1}} \cdot \cdots Y$$

and that, in view of 33.8(ii) and the fact that $a_{i-1} f_i = g_i a_i$, these last two zigzags are in the same class in the sense of 33.8.

34. A colimit description of the hom-sets of the homotopy category

In this section we deduce from the description of the hom-sets of the homotopy category $\operatorname{Ho} \boldsymbol{C}$ of a homotopical category $\boldsymbol{C}$ in which, for every two objects $X, Y \in \boldsymbol{C}$, the hom-set $\operatorname{Ho} \boldsymbol{C}(X, Y)$ is described as a set of equivalence classes of restricted zigzags in $\boldsymbol{C}$ from X to Y, where each such class contains zigzags of *countably* many different types (i.e. numbers of forward and backward maps and the order in which they appear), a description of $\operatorname{Ho} \boldsymbol{C}(X, Y)$ as a *colimit of a diagram of sets*, indexed by the category $\boldsymbol{T}$ of these types, in which each set consists of equivalence classes of zigzags of only *one* type.

To do this we start with giving a precise definition of

34.1. Types of zigzags. The **type** of a zigzag in a category $\boldsymbol{C}$ from an object X to an object Y (33.8)

$$X = C_0 \xrightarrow{f_1} C_1 \xrightarrow{} \cdots \xrightarrow{f_n} C_n = Y \qquad (n \geq 0)$$

will be the pair $T = (T_+, T_-)$ of complementary subsets of the set of integers $\{1, \ldots, n\}$ such that $i \in T_+$ whenever f_i is forward and $i \in T_-$ otherwise.

These types can be considered as objects of a **category of types $\boldsymbol{T}$** which has, for every two types (T_+, T_-) and (T'_+, T'_-) of length n and n' respectively, as maps $t \colon (T_+, T_-) \to (T'_+, T'_-)$ the weakly monotone functions $t \colon \{1, \ldots, n\} \to \{1, \ldots, n'\}$ such that

$$t(T_+) \subset T'_+ \qquad \text{and} \qquad t(T_-) \subset T'_-.$$

With these types of zigzags one can associate

34.2. Arrow categories. Given a homotopical category $\boldsymbol{C}$, a pair of objects $X, Y \in \boldsymbol{C}$ and a type T of length n (34.1), the associated n-**arrow category** will be the category which has as objects the **restricted zigzags** in $\boldsymbol{C}$ from X to Y (i.e. zigzags in which the backward maps are weak equivalences) of type T and has as maps the commutative diagrams in $\boldsymbol{C}$ of the form

$$\begin{array}{ccccc} X & \longrightarrow \cdot \cdots \cdot \longrightarrow & Y \\ {\scriptstyle 1}\downarrow & \downarrow \quad \downarrow & \downarrow {\scriptstyle 1} \\ X & \longrightarrow \cdot \cdots \cdot \longrightarrow & Y \end{array}$$

in which the top row and the bottom row are restricted zigzags in $\boldsymbol{C}$ of type T and the vertical maps are *weak equivalences*. This n-arrow category will be denoted by
$$\boldsymbol{C}^T(X,Y)$$

Next we note that from these arrow categories one can form

34.3. $\boldsymbol{T}$-diagrams of arrow categories. Given a homotopical category $\boldsymbol{C}$ and a pair of objects $X, Y \in \boldsymbol{C}$, one can form a **$\boldsymbol{T}$-diagram of arrow categories**
$$\boldsymbol{C}^{(\boldsymbol{T})}(X,Y)$$
which assigns to every object $T \in \boldsymbol{T}$ (34.1) the category (34.2)
$$\boldsymbol{C}^T(X,Y)$$
and to every map $t\colon T \to T' \in \boldsymbol{T}$ the functor
$$t_*\colon \boldsymbol{C}^T(X,Y) \to \boldsymbol{C}^{T'}(X,Y)$$
which sends a zigzag of type T
$$X = C_0 \xrightarrow{f_1} C_1 \underline{} \cdot \ \cdots \ \cdot \xrightarrow{f_n} C_n = Y$$
to the zigzag of type T'
$$X = C'_0 \xrightarrow{f'_1} C'_1 \underline{} \cdot \ \cdots \ \cdot \xrightarrow{f'_n} C'_{n'} = Y$$
in which each f'_j ($1 \le j \le n'$) is the composition of the f_i with $ti = j$ or, if no such i exists, the appropriate identity map.

Similarly, for every subcategory $\boldsymbol{T}' \subset \boldsymbol{T}$ the restriction of the above $\boldsymbol{T}$-diagram to $\boldsymbol{T}'$ yields a **$\boldsymbol{T}'$-diagram of arrow categories** which we will denote by
$$\boldsymbol{C}^{(\boldsymbol{T}')}(X,Y)$$

We also need the notion of

34.4. Connected components of a category. A **connected component** of a category $\boldsymbol{D}$ is a *maximal* subcategory $\boldsymbol{D}_0 \subset \boldsymbol{D}$ with the property that every pair of objects in $\boldsymbol{D}_0$ can be connected by a zigzag of maps in $\boldsymbol{D}_0$. We will denote by $\pi_0 \boldsymbol{D}$ the resulting set of the equivalence classes of the objects of $\boldsymbol{D}$, where two objects will be in the same class when they are in the same connected component, and denote, for a functor $c\colon \boldsymbol{D} \to \boldsymbol{E}$, by $\pi_0 c\colon \pi_0 \boldsymbol{D} \to \pi_0 \boldsymbol{E}$ the function induced by c.

Now we can finally formulate the promised

34.5. Colimit description of the hom-sets of the homotopy category. Let $\boldsymbol{C}$ be a homotopical category. Then, for every pair of objects $X, Y \in \boldsymbol{C}$ (34.3)
$$\operatorname{Ho}\boldsymbol{C}(X,Y) = \operatorname{colim}^{\boldsymbol{T}} \pi_0 \boldsymbol{C}^{(\boldsymbol{T})}(X,Y) \ .$$

Proof. This follows from the observation that in view of the first two commutative diagrams in the proof of 33.10 the functor π_0 takes care of the relations imposed by 33.8(iii) and that the functors between the arrow categories take exactly care of the relations imposed by 33.8(i) and (ii).

35. A Grothendieck construction

As a variation on the above colimit description of the hom-sets of the homotopy category of a homotopical category $\boldsymbol{C}$, we

(i) obtain, for every pair of objects $X, Y \in \boldsymbol{C}$, from the $\boldsymbol{T}$-diagram of arrow categories $\boldsymbol{C}^{(\boldsymbol{T})}(X,Y)$ a kind of homotopy colimit which is a single category $\operatorname{Gr} \boldsymbol{C}^{(\boldsymbol{T})}(X,Y)$ (called its *Grothendieck construction*) which has as its objects all the restricted zigzags in $\boldsymbol{C}$ from X to Y and of which any two objects are in the same connected component iff (34.4) they represent the same map $X \to Y \in \operatorname{Ho} \boldsymbol{C}$, and

(ii) combine all these Grothendieck constructions into a single category

$$\operatorname{Gr} \boldsymbol{C}^{(\boldsymbol{T})}$$

which has the same objects as $\boldsymbol{C}$ and is *enriched over* **CAT** (32.3) and which has the property that its "category of connected components" is exactly the category $\operatorname{Ho} \boldsymbol{C}$.

(iii) We also note that this enriched category is closely related to the *(simplicial) hammock localization* of [**DK80a**].

We thus start with

35.1. Grothendieck construction. We recall from [**Tho79**] that, given a homotopical category $\boldsymbol{C}$ and a pair of objects $X, Y \in \boldsymbol{C}$, the **Grothendieck construction** on the $\boldsymbol{T}$-diagram of categories $\boldsymbol{C}^{(\boldsymbol{T})}(X,Y)$ (34.3) is the category

$$\operatorname{Gr} \boldsymbol{C}^{(\boldsymbol{T})}(X,Y)$$

which has (34.2 and 34.3)

(i) as objects the restricted zigzags in $\boldsymbol{C}$ from X to Y, i.e. the pairs (T,Z) consisting of the objects

$$T \in \boldsymbol{T} \qquad \text{and} \qquad Z \in \boldsymbol{C}^T(X,Y)$$

and

(ii) for every two such objects (T', Z') and (T'', Z''), as maps $(T', Z') \to (T'', Z'')$ the pairs (t, z) consisting of maps

$$t \colon T' \to T'' \in \boldsymbol{T} \qquad \text{and} \qquad z \colon t_* Z' \to Z'' \in \boldsymbol{C}^{T''}(X,Y)$$

and in which

(iii) for every two composable maps (t, z) and (t', z'), their composition is defined by the formula

$$(t', z')(t, z) = \bigl(t't, z'(t_* z)\bigr)$$

and note that, by essentially the same argument as was used in the proof of 34.5, this definition implies that

35.2. Proposition. *For every homotopical category $\boldsymbol{C}$ and pair of objects $X, Y \in \boldsymbol{C}$,* (34.5)

$$\operatorname{Ho} \boldsymbol{C}(X,Y) = \operatorname{colim}^{\boldsymbol{T}} \pi_0 \boldsymbol{C}^{(\boldsymbol{T})}(X,Y) = \pi_0 \operatorname{Gr} \boldsymbol{C}^{(\boldsymbol{T})}(X,Y) \ .$$

Next we note that these Grothendieck constructions give rise to what we will call the

35.3. Grothendieck enrichment. Given a homotopical category C with category of weak equivalences W, its **Grothendieck enrichment** will be the category $\operatorname{Gr} C^{(T)}$ *enriched* over **CAT** (32.3) which

(i) has the same objects as C,
(ii) has, for every two objects $X, Y \in C$, as its hom-category the category $\operatorname{Gr} C^{(T)}(X, Y)$, and
(iii) has, for every three objects X_1, X_2 and $X_3 \in C$ as its composition functor

$$\operatorname{Gr} C^{(T)}(X_1, X_2) \times \operatorname{Gr} C^{(T)}(X_2, X_3) \longrightarrow \operatorname{Gr} C^{(T)}(X_1, X_3)$$

the functor induced by the compositions of the zigzags involved.

If $\pi_0 \operatorname{Gr} C^{(T)}$ denotes the category obtained from $\operatorname{Gr} C^{(T)}$ by applying the functor π_0 (34.4) to each of its hom-categories, then 34.5 and 35.2 imply the following

35.4. Grothendieck description of the homotopy category. *For every homotopical category C*

$$\operatorname{Ho} C = \pi_0 \operatorname{Gr} C^{(T)} \ .$$

We end with a brief discussion of the connection between these Grothendieck enrichments and simplicial localizations and for this we need a few

35.5. Simplicial preliminaries. We need the following definitions and results.

(i) a **simplicial set** will be a functor $A \colon \Delta^{\mathrm{op}} \to \mathbf{SET}$, where **SET** denotes the $\mathcal{U}^+$-category (32.1 and 32.2) of sets (32.3) and Δ is the category which has as its objects the categories $[n]$ ($n \geq 0$) (22.10) and as maps all the functors between them, and the elements of the set $A[n]$ ($n \geq 0$) will be referred to as the n-**simplices** of A,
(ii) every category X then gives, in a functorial manner, rise to a simplicial set $\mathrm{N} X$ (called its **nerve** or **classifying space**) which is the functor $\Delta^{\mathrm{op}} \to \mathbf{SET}$ which sends $[n]$ ($n \geq 0$) to the set of the functors $[n] \to X$, and
(iii) the $\mathcal{U}^+$-category of simplicial sets $\mathbf{SET}^{\Delta^{\mathrm{op}}}$ admits a *model structure* (9.1) (the details of which can be found in [**GJ99**], [**Hir03**] and [**Hov99**]) in which
(iii)' the *cofibrations* are the objectwise monomorphisms, and
(iii)'' for every pair of functors $f \colon X \to Y$ and $g \colon Y \to X$ for which the compositions gf and fg can be connected by a natural transformation to the identity functors of respectively X and Y, the induced maps between nerves $\mathrm{N} f \colon \mathrm{N} X \to \mathrm{N} Y$ and $\mathrm{N} g \colon \mathrm{N} Y \to \mathrm{N} X$ are *weak equivalences*.

Now we can start dealing with

35.6. The connection with simplicial localizations. Given a homotopical category C, its Grothendieck enrichment (35.3) gives rise to a **simplicial category**, i.e. a category enriched over simplicial sets (35.5)
$$\mathrm{N}\,\mathrm{Gr}\,\boldsymbol{C}^{(\boldsymbol{T})}$$
which has the same objects as C and which, for every pair of objects $X, Y \in C$, has the simplicial set
$$\mathrm{N}\,\mathrm{Gr}\,\boldsymbol{C}^{(\boldsymbol{T})}(X,Y)$$
as its simplicial hom-set. This simplicial category is closely related to the **simplicial hammock localization** of C, i.e. the simplicial category $L^H C$ which has the same objects as C and in which, for every pair of objects $X, Y \in C$, the simplicial hom-set can be described as [**DK80a**, 5.5]
$$L^H \boldsymbol{C}(X,Y) = \operatorname{colim}^{\boldsymbol{T}}\mathrm{N}\boldsymbol{C}^{(\boldsymbol{T})}(X,Y)\ .$$

These two simplicial categories then are related by the following

35.7. Proposition. *Let C be a homotopical category. Then the map of simplicial categories (35.6)*
$$p\colon \mathrm{N}\,\mathrm{Gr}\,\boldsymbol{C}^{(\boldsymbol{T})} \to L^H \boldsymbol{C}$$
which is the identity on the objects and which, for every pair of object $X, Y \in C$, consists of the map
$$p(X,Y)\colon \mathrm{N}\,\mathrm{Gr}\,\boldsymbol{C}^{(\boldsymbol{T})}(X,Y) \longrightarrow \operatorname{colim}^{\boldsymbol{T}} \mathrm{N}\boldsymbol{C}^{(\boldsymbol{T})}(X,Y)$$
which sends an n-simplex of $\mathrm{N}\,\mathrm{Gr}\,\boldsymbol{C}^{(\boldsymbol{T})}(X,Y)$, i.e. a sequence of maps in the category $\mathrm{Gr}\,\boldsymbol{C}^{(\boldsymbol{T})}(X,Y)$ (35.1)
$$(T^0, Z^0) \xrightarrow{(t^1, z^1)} (T^1, Z^1) \longrightarrow \cdots \longrightarrow (T^{n-1}, Z^{n-1}) \xrightarrow{(t^n, z^n)} (T^n, Z^n)$$
to the n-simplex of $\operatorname{colim}^{\boldsymbol{T}} \mathrm{N}\boldsymbol{C}^{(\boldsymbol{T})}(X,Y)$ which contains the sequence of maps in $\boldsymbol{C}^{T^n}(X,Y)$
$$(t^n \cdots t^1)_* Z^0 \xrightarrow{(t^n \cdots t^2)_* z^1} (t^n \cdots t^2)_* Z^1 \longrightarrow \cdots \longrightarrow t^n_* Z^{n-1} \xrightarrow{z^n} Z^n$$
is a weak equivalence in the sense that, for every pair of objects $(X, Y) \in C$, the map $p(X, Y)$ is a weak equivalence of simplicial sets (35.5).

To prove this we need the following

35.8. Colimit description of the Grothendieck construction. Given a homotopical category C and objects $X, Y \in C$, let $r\boldsymbol{C}^{(\boldsymbol{T})}(X,Y)$ denote the $\boldsymbol{T}$-diagram of categories which associates

(i) with every object $T \in \boldsymbol{T}$ the category $r\boldsymbol{C}^T(X,Y)$ which has as its objects the pairs $((\overline{T}, \overline{Z}), t)$ consisting of an object $(\overline{T}, \overline{Z}) \in \mathrm{Gr}\,\boldsymbol{C}^{(\boldsymbol{T})}(X,Y)$ and a map $t\colon \overline{T} \to T \in \boldsymbol{T}$ and as its maps $((\overline{T}, \overline{Z}), t) \to ((\overline{T}', \overline{Z}'), t')$ the maps
$$(\bar{t}, \bar{z})\colon (\overline{T}, \overline{Z}) \longrightarrow (\overline{T}', \overline{Z}') \in \mathrm{Gr}\,\boldsymbol{C}^{(\boldsymbol{T})}(X,Y)$$
for which $t'\bar{t} = t$

and which associates

(ii) with every map $t_0\colon T \to T' \in \boldsymbol{T}$ the functor $r\boldsymbol{C}^T(X,Y) \to r\boldsymbol{C}^{T'}(X,Y)$ which sends an object $\bigl((\overline{T},\overline{Z}),t\bigr)$ to the object $\bigl((\overline{T},\overline{Z}),t_0 t\bigr)$.

A straightforward calculation (which will be left to the reader) then yields that

(iii) *the forgetful functors*
$$r\boldsymbol{C}^T(X,Y) \longrightarrow \operatorname{Gr} \boldsymbol{C}^{(\boldsymbol{T})}(X,Y) \qquad (T \in \boldsymbol{T})$$
induce an isomorphism
$$\operatorname{colim}^{\boldsymbol{T}} r\boldsymbol{C}^{(\boldsymbol{T})}(X,Y) \approx \operatorname{Gr} \boldsymbol{C}^{(\boldsymbol{T})}(X,Y)$$

and that moreover

(iv) *the categories $r\boldsymbol{C}^T(X,Y)$ and $\boldsymbol{C}^T(X,Y)$ ($T \in \boldsymbol{T}$) are connected by*

(iv)′ *a functor*
$$s\colon r\boldsymbol{C}^T(X,Y) \longrightarrow \boldsymbol{C}^T(X,Y)$$
which sends
- *an object $\bigl((\overline{T},\overline{Z}),t\bigr)$ to the object $t_*\overline{Z}$ and*
- *a map $(\bar{t},\bar{z})\colon \bigl((\overline{T},\overline{Z}),t\bigr) \to \bigl((\overline{T}',\overline{Z}'),t'\bigr)$ to the map $t'_*\bar{z}\colon t_*\overline{Z} = t'_*\bar{t}\overline{Z} \to t'_*\overline{Z}'$,*

and which is natural in T, and

(iv)″ *a functor*
$$i\colon \boldsymbol{C}^T(X,Y) \longrightarrow r\boldsymbol{C}^T(X,Y)$$
which sends an object Z to the object $\bigl((T,Z),1_T\bigr)$ and a map $z\colon Z \to Z'$ to the map $(1_T,z)\colon \bigl((T,Z),1_T\bigr) \to \bigl((T,Z'),1^T\bigr)$ and which has the properties that the composition si is the identity map of $\boldsymbol{C}^T(X,Y)$ and that the function which assigns to every object $\bigl((\overline{T},\overline{Z}),t\bigr) \in r\boldsymbol{C}^T(X,Y)$ the map $(t,1_{t_*\overline{Z}})\colon \bigl((\overline{T},\overline{Z}),t\bigr) \to \bigl((T,t_*\overline{Z}),1_T\bigr)$ is a natural transformation from the identity map of $r\boldsymbol{C}^T(X,Y)$ to the composition is.

It thus remains to give an outline of the

35.9. Proof of 35.7. To prove 35.7 one first notes that the map $p(X,Y)$ admits a factorization

$$\begin{array}{ccc}
\operatorname{N}\operatorname{Gr}\boldsymbol{C}^{(\boldsymbol{T})}(X,Y) & \xrightarrow{p(X,Y)} & \operatorname{colim}^{\boldsymbol{T}} \operatorname{N}\boldsymbol{C}^{(\boldsymbol{T})}(X,Y) \\
{\scriptstyle j}\downarrow\approx & & \uparrow{\scriptstyle \operatorname{colim}^{\boldsymbol{T}} \operatorname{N}s} \\
\operatorname{N}\operatorname{colim}^{\boldsymbol{T}} r\boldsymbol{C}^{(\boldsymbol{T})}(X,Y) & \xrightarrow{k} & \operatorname{colim}^{\boldsymbol{T}} \operatorname{N}r\boldsymbol{C}^{(\boldsymbol{T})}(X,Y)
\end{array}$$

in which j is the isomorphism induced by the inverse of the isomorphism of 35.8(iii), s is as in 35.8(iv)′ and k is the map such that the composition kj sends, for every integer $n \geq 0$, an n-simplex of $\operatorname{N}\operatorname{Gr}\boldsymbol{C}^{(\boldsymbol{T})}(X,Y)$, i.e. a sequence of maps in $\operatorname{Gr}\boldsymbol{C}^{(\boldsymbol{T})}(X,Y)$

$$(T_0,Z_0) \xrightarrow{(t^1,z^1)} \cdots \xrightarrow{(t^{n-1},z^{n-1})} (T^{n-1},Z^{n-1}) \xrightarrow{(t^n,z^n)} (T^n,Z^n)$$

to the n-simplex of $\operatorname{colim}^{\boldsymbol{T}} \operatorname{N}r\boldsymbol{C}^{(\boldsymbol{T})}(X,Y)$ which contains the n-simplex

$$\bigl((T^0,Z^0),t^n\cdots t^1\bigr) \xrightarrow{(t^1,z^1)} \cdots \longrightarrow \bigl((T^{n-1},Z^{n-1}),t^n\bigr) \xrightarrow{(t^n,z^n)} \bigl((T^n,Z^n),1_{T^n}\bigr)$$

of $\operatorname{N}r\boldsymbol{C}^{T^n}(X,Y)$.

One then readily verifies that kj and hence k is an isomorphism and it thus remains to show that the map $\operatorname{colim}^T \mathrm{N}s$ is a weak equivalence. To do this one notes that

(i) the category $\boldsymbol{T}$ can be considered as a *Reedy category* with *fibrant constants* (22.1 and 22.8) in which the required subcategories $\overrightarrow{\boldsymbol{T}}$ and $\overleftarrow{\boldsymbol{T}}$ consist of the maps which are respectively 1-1 and onto, and that

(ii) the $\boldsymbol{T}$-diagrams of simplicial sets

$$\mathrm{N}\boldsymbol{C}^{(\boldsymbol{T})}(X,Y) \quad \text{and} \quad \mathrm{N}r\boldsymbol{C}^{(\boldsymbol{T})}(X,Y)$$

are, in view of 35.5(iii)′ both *Reedy cofibrant diagrams* (10.2, 22.2 and 22.6),

and the desired result now follows from 22.7 and the observation that, in view of 33.5(iii)″ and 35.8(iv)″ the maps

$$\mathrm{N}s\colon \mathrm{N}r\boldsymbol{C}^T(X,Y) \longrightarrow \mathrm{N}\boldsymbol{C}^T(X,Y) \qquad (T\in\boldsymbol{T})$$

are weak equivalences.

36. 3-arrow calculi

Next we use the description (34.5) of the hom-sets of the homotopy category of a homotopical category as a colimit of a diagram of sets, indexed by the category of types $\boldsymbol{T}$, in which each set consists of equivalence classes of restricted zigzags of only one type, to show that

(i) if $\boldsymbol{C}$ admits a so-called 3-*arrow calculus*, then these hom-sets can be described in terms of zigzags in $\boldsymbol{C}$ of type

$$\cdot \longleftarrow \cdot \longrightarrow \cdot \longleftarrow \cdot$$

only.

Moreover we deduce from this (for the first time using the full strength of the "two out of six" property (33.1) and not merely the "two out of three" property) the somewhat unexpected result that

(ii) if $\boldsymbol{C}$ admits a 3-arrow calculus, then $\boldsymbol{C}$ is *saturated*, i.e. (33.9) a map in $\boldsymbol{C}$ is a weak equivalence iff its image in $\operatorname{Ho}\boldsymbol{C}$ is an isomorphism.

This last result turns out to be of use in chapters VII and VIII where we obtain sufficient conditions for the composability of approximations (in §42) and of homotopy colimit and limit functors (in §47) and for the homotopical cocompleteness and completeness of homotopical categories (in §49), which are considerably simpler, and therefore easier to verify, when the homotopical categories involved are saturated.

We start (motivated by some of the properties of the subcategories of the trivial cofibrations and the trivial fibrations in a model category (see 11.4)) with defining

36.1. 3-arrow calculi. Given a homotopical category $\boldsymbol{C}$ with category of weak equivalences $\boldsymbol{W}$ (33.1), $\boldsymbol{C}$ is said to admit a **3-arrow calculus** $\{U,V\}$ if there exist subcategories $\boldsymbol{U},\boldsymbol{V}\subset\boldsymbol{W}$ such that

(i) for every zigzag $A' \xleftarrow{u} A \xrightarrow{f} B$ in $\boldsymbol{C}$ which $u \in \boldsymbol{U}$, there exists a *functorial* zigzag $A' \xrightarrow{f'} B' \xleftarrow{u'} B$ in $\boldsymbol{C}$ with $u' \in \boldsymbol{U}$ such that

$$u'f = f'u \text{ and}$$

u' is an isomorphism whenever u is

(e.g. if $\boldsymbol{C}$ is closed under pushouts and every pushout of a map in $\boldsymbol{U}$ is again in $\boldsymbol{U}$),

(ii) for every zigzag $X \xrightarrow{g} Y \xleftarrow{v} Y'$ in $\boldsymbol{C}$ with $v \in \boldsymbol{V}$, there exists a *functorial* zigzag $X \xleftarrow{v'} X' \xrightarrow{g'} Y$ in $\boldsymbol{C}$ with $v' \in \boldsymbol{V}$ such that

$$gv' = vg' \text{ and}$$

v' is an isomorphism whenever v is

(e.g. if $\boldsymbol{C}$ is closed under pullbacks and every pullback of a map in $\boldsymbol{V}$ is again in $\boldsymbol{V}$), and

(iii) every map $w \in \boldsymbol{W}$ admits a *functorial* factorization $w = vu$ with $u \in \boldsymbol{U}$ and $v \in \boldsymbol{V}$.

Note that (iii) implies that $\boldsymbol{U}$ and $\boldsymbol{V}$ contain all the objects of $\boldsymbol{W}$ and hence of $\boldsymbol{C}$.

Another immediate consequence of this definition is that

(iv) *for every homotopical category $\boldsymbol{D}$, the homotopical diagram categories* (33.2)

$$\boldsymbol{C}^{\boldsymbol{D}} \quad \text{and} \quad (\boldsymbol{C}^{\boldsymbol{D}})_{\mathrm{w}}$$

inherit from $\boldsymbol{C}$ a 3-arrow calculus in which the desired subcategories consist of the natural transformations which send the objects of $\boldsymbol{D}$ to maps in $\boldsymbol{U}$ and $\boldsymbol{V}$ respectively.

Similarly one has

36.2. Proposition. *Let $\boldsymbol{C}$ be a homotopical category which admits a 3-arrow calculus $\{\boldsymbol{U}, \boldsymbol{V}\}$ (36.1). Then, for every pair of objects $X, Y \in \boldsymbol{C}$, the 3-arrow category* (34.2)

$$\boldsymbol{C}^{(\{2\},\{1,3\})}(X, Y)$$

and its full subcategory

$$\boldsymbol{C}^{(\{2\},\{1,3\})}_{\{\boldsymbol{U},\boldsymbol{V}\}}(X, Y)$$

spanned by the zigzags of the form

$$X \xleftarrow{v} \cdot \longrightarrow \cdot \xleftarrow{u} Y$$

in which $u \in \boldsymbol{U}$ and $v \in \boldsymbol{V}$, inherit from $\boldsymbol{C}$ a 3-arrow calculus of which

(i) *the desired subcategories consist of the commutative diagrams in $\boldsymbol{C}$ of the form*

$$\begin{array}{ccccc} X & \leftarrow & \cdot & \rightarrow \cdot \leftarrow & Y \\ {\scriptstyle 1}\downarrow & & \downarrow & \downarrow & \downarrow {\scriptstyle 1} \\ X & \leftarrow & \cdot & \rightarrow \cdot \leftarrow & Y \end{array}$$

in which the vertical maps are respectively in $\boldsymbol{U}$ and in $\boldsymbol{V}$, and

(ii) *the functorial factorization consists of the commutative diagrams in $\boldsymbol{C}$ of the form*

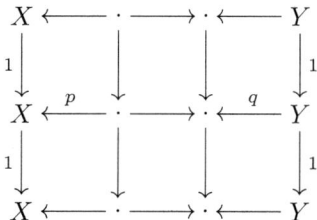

in which the squares in the middle are produced by the given functorial factorization and the maps p and q are the unique ones which make the other squares commutative.

Now we formulate the promised

36.3. 3-arrow description of the hom-sets of the homotopy category.
Let $\boldsymbol{C}$ be a homotopical category which admits a 3-arrow calculus $\{\boldsymbol{U}, \boldsymbol{V}\}$ (36.1). Then, for every pair of objects $X, Y \in \boldsymbol{C}$,

(i) *the map (34.1–34.5)*

$$\pi_0 \boldsymbol{C}^{(\{2\},\{1,3\})}(X,Y) \xrightarrow{\text{incl.}_*} \operatorname{colim}^{\boldsymbol{T}} \pi_0 \boldsymbol{C}^{(\boldsymbol{T})}(X,Y) = \operatorname{Ho} \boldsymbol{C}(X,Y)$$

induced by the inclusion $(\{2\},\{1,3\}) \in \boldsymbol{T}$ and the map

$$\pi_0 \boldsymbol{C}^{(\{2\},\{1,3\})}_{\{\boldsymbol{U},\boldsymbol{V}\}}(X,Y) \xrightarrow{\text{incl.}_*} \pi_0 \boldsymbol{C}^{(\{2\},\{1,3\})}(X,Y)$$

induced by the inclusion of 36.2, are both isomorphisms.

Moreover, in view of (i) and 36.2

(ii) *two objects of $\boldsymbol{C}^{(\{2\},\{1,3\})}(X,Y)$ are in the same connected component iff they are the top row and the bottom row in a commutative diagram in $\boldsymbol{C}$ of the form*

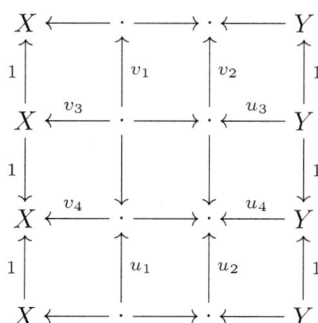

in which u_1 and u_2 are in $\boldsymbol{U}$ and v_1 and v_2 are in $\boldsymbol{V}$, and

(iii) *two objects of $\boldsymbol{C}^{(\{2\},\{1,3\})}_{\{\boldsymbol{U},\boldsymbol{V}\}}(X,Y)$ are in the same connected component iff they are the top row and the bottom row in such a diagram in which all the u's are in $\boldsymbol{U}$ and all the v's are in $\boldsymbol{V}$.*

110   VI. HOMOTOPICAL CATEGORIES AND HOMOTOPICAL FUNCTORS

Before proving this result we first use it to obtain the earlier mentioned

36.4. Saturation proposition. *Let C be a homotopical category with localization functor $\gamma\colon C \to \operatorname{Ho} C$ (33.9), which admits a 3-arrow calculus (36.1). Then C is saturated, i.e. (33.9) a map $f \in C$ is a weak equivalence iff the map $\gamma f \in \operatorname{Ho} C$ is an isomorphism and moreover (33.9(v) or 36.1(iv)) so are, for every homotopical category D, the homotopical functor categories (33.2)*

$$\operatorname{Fun}(D, C) \quad \text{and} \quad \operatorname{Fun}_w(D, C).$$

Proof. Let $\{U, V\}$ be a 3-arrow calculus on C, let $f\colon A \to B \in C$ be a map such that $\gamma f \in \operatorname{Ho} C$ is an isomorphism and let (36.3(i))

$$B \xleftarrow{v} B' \xrightarrow{g} A' \xleftarrow{u} A$$

be a zigzag in C with $u \in U$ and $v \in V$ which represents $(\gamma f)^{-1}$. Then one can form the commutative diagram

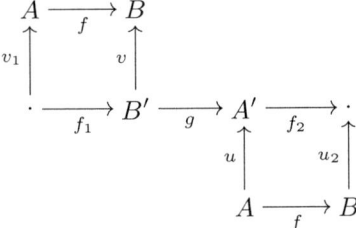

in which the squares are as in 36.1(ii) and (i). In this diagram the zigzags

$$\cdot \xrightarrow{f} \cdot \xleftarrow{v} \cdot \xrightarrow{g} \cdot \xleftarrow{u} \cdot \quad \text{and} \quad \cdot \xleftarrow{v} \cdot \xrightarrow{g} \cdot \xleftarrow{u} \cdot \xrightarrow{f} \cdot$$

represent identity maps in $\operatorname{Ho} C$ and hence so do the zigzags

$$\cdot \xleftarrow{v_1} \cdot \xrightarrow{gf_1} \cdot \xleftarrow{u} \cdot \quad \text{and} \quad \cdot \xleftarrow{v} \cdot \xrightarrow{f_2 g} \cdot \xleftarrow{u_2} \cdot \;.$$

This implies, in view of 36.3(ii) and the "two out of three" property (33.1) that gf_1 and $f_2 g$ are weak equivalences and so are therefore *in view of the "two out of six" property* (33.1), the maps g, f_1 and f_2 and another application of the two out of three property now yields the desired result that f is a weak equivalence.

It thus remains to give a

36.5. Proof of 36.3. That the second map in 36.3(i) is an isomorphism follows from the fact that (36.1) every restricted zigzag in C of the form $X \leftarrow \cdot \to \cdot \leftarrow Y$ can, in a functorial manner, be embedded as the top row in a zigzag in $C^{(\{2\},\{1,3\})}(X, Y)$ of the form

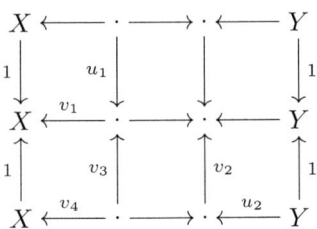

36. 3-ARROW CALCULI

in which the u's are in $\boldsymbol{U}$ and the v's are in $\boldsymbol{V}$ and the squares in the middle are as in 36.1(i) and (ii) respectively.

That the first map in 36.3(i) is also an isomorphism, we will show by proving that, in the commutative diagram (34.3)

$$\begin{array}{ccc}
\pi_0 \boldsymbol{C}^{(\{2\},\{1,3\})}(X,Y) & \xrightarrow{\mathrm{incl.}_*} & \mathrm{colim}^{\boldsymbol{T}}\, \pi_0 \boldsymbol{C}^{(\boldsymbol{T})}(X,Y) \\
{\scriptstyle p}\downarrow & {\scriptstyle q}\nearrow & \downarrow{\scriptstyle r} \\
\mathrm{colim}^{E\boldsymbol{T}}\, \pi_0 \boldsymbol{C}^{(E\boldsymbol{T})}(X,Y) & \xrightarrow{\mathrm{incl.}_*} & \mathrm{colim}^{\boldsymbol{T}}\, \pi_0 \boldsymbol{C}^{(\boldsymbol{T})}(X,Y)
\end{array}$$

in which $E\boldsymbol{T} \subset \boldsymbol{T}$ denotes the image of $\boldsymbol{T}$ in $\boldsymbol{T}$ under the functor $E\colon \boldsymbol{T} \to \boldsymbol{T}$ given by the formula

$$E\big(\{i_1,\cdots,i_a\},\{j_1,\cdots j_b\}\big) = \big(\{i_1+1,\cdots,i_a+1\},\{1,j_1+1,\cdots,j_b+1,a+b+2\}\big)$$

and q is the map which sends, for every restricted zigzag in $\boldsymbol{C}$ from X to Y, the element containing it to the element containing the zigzag obtained from it by adding a backward identity map on both sides, the maps p and r (and hence the other three maps) are isomorphisms.

That the map r is an isomorphism follows readily from 33.8(i), but in order to be able to prove that the map p is also an isomorphism we first note that p admits a factorization

$$\pi_0 \boldsymbol{C}^{(\{2\},\{1,3\})}(X,Y) \xrightarrow{p'} \mathrm{colim}^{FE\boldsymbol{T}}\, \pi_0 \boldsymbol{C}^{(FE\boldsymbol{T})}(X,Y)$$

$$\xrightarrow{\mathrm{incl.}_*} \mathrm{colim}^{E\boldsymbol{T}}\, \pi_0 \boldsymbol{C}^{(E\boldsymbol{T})}(X,Y)$$

in which $FE\boldsymbol{T} \subset E\boldsymbol{T}$ is the image of $E\boldsymbol{T}$ in $E\boldsymbol{T}$ under the functor $F\colon E\boldsymbol{T} \to E\boldsymbol{T}$ given by the formula

$$F\big(\{i_1,\cdots,i_c\},\{1,j_1,\cdots,j_d,c+d+2\}\big) = \big(\{2,\cdots,c+1\},\{1,c+2\}\big)\ .$$

That the second of these maps is an isomorphism then follows from the observations that the functor F is the identity on $FE\boldsymbol{T}$ and comes with a natural transformation $f\colon F \to 1_{E\boldsymbol{T}}$ which sends each object $T \in E\boldsymbol{T}$ to the *unique* monomorphism $FT \to T \in E\boldsymbol{T}$ and that, for every object $T \in E\boldsymbol{T}$ the map (34.3)

$$\pi_0\big(fT\big)_*\colon \pi_0 \boldsymbol{C}^{FT}(X,Y) \longrightarrow \pi_0 \boldsymbol{C}^{T}(X,Y)$$

is an isomorphism, which last observation can be verified by iterated application of the fact that, in view of the presence of the 3-arrow calculus $\{\boldsymbol{U},\boldsymbol{V}\}$ (36.1), one can, in a *functorial* manner, embed every restricted zigzag in $\boldsymbol{C}$ of the form

$$\cdot \leftarrow \cdot \ \cdots \ \cdot \leftarrow \cdot \ \cdots \ \cdot \leftarrow \cdot$$

in which each $\cdots$ indicates a possibly empty sequence of *forward* maps, in a commutative diagram in $\boldsymbol{C}$ of the form

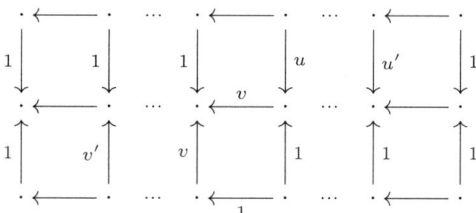

in which $u, u' \in \boldsymbol{U}$, $v, v' \in \boldsymbol{V}$ and u' and v' are obtained from u and v by iterated applications of 36.1(i) and 36.1(ii) respectively.

To finally prove that the map p' is also an isomorphism we note that, for every restricted zigzag of the form

$$X \xleftarrow{w_1} \cdot \xrightarrow{z} \cdot \xleftarrow{w_2} Y$$

p' sends the element containing it to the element containing the zigzag

$$X \xleftarrow{1} \cdot \xleftarrow{w_1} \cdot \xrightarrow{z} \cdot \xleftarrow{w_2} \cdot \xleftarrow{1} Y$$

which, in view of the commutativity of the diagram

$$X \xleftarrow{1} \cdot \xleftarrow{w_1} \cdot \xrightarrow{z} \cdot \xleftarrow{w_2} \cdot \xleftarrow{1} Y$$

also contains its bottom row and hence the original zigzag

$$X \xleftarrow{w_1} \cdot \xrightarrow{z} \cdot \xleftarrow{w_2} Y$$

and that therefore the map p' is exactly the map induced by the inclusion of $(\{2\}, \{1,3\})$ in $FE\boldsymbol{T}$, and the desired result now follows immediately from the observation that $(\{2\}, \{1,3\})$ is the *terminal* object of $FE\boldsymbol{T}$.

37. Homotopical uniqueness

In classical category theory one often notes that certain objects in a category are *canonically isomorphic* or *unique up to a unique isomorphism* in the sense that, for any two of these objects X and Y, there is *exactly one* map $X \to Y$ in the category and that this unique map is an *isomorphism* (for instance (see 37.3 and 37.4) because they have a common *universal property*) and our aim in this section is to describe a homotopical version of this phenomenon.

We start with a slight reformulation of the above mentioned classical results.

37.1. Categorical uniqueness. Given a category $\boldsymbol{Y}$ and a non-empty set I (32.3), we will say that objects $Y_i \in \boldsymbol{Y}$ ($i \in I$) are **canonically isomorphic** or **categorically unique** if

   (i) the *full* subcategory of $\boldsymbol{Y}$ spanned by these objects

or equivalently

   (ii) the **categorically full** subcategory of $\boldsymbol{Y}$ spanned by these objects, i.e. the full subcategory spanned by these objects and all isomorphic ones

is *categorically contractible* in the following sense.

37.2. Categorically contractible categories. A category G will be called **categorically contractible** if

a) G is a *non-empty groupoid* in which there is *exactly one* isomorphism between any two objects,

or equivalently

b) the unique functor $G \to [0]$ (where $[0]$ denotes the category which consists of only one object 0 and its identity map) is *an equivalence of categories*.

One then readily verifies that this is equivalent to requiring that G be a *non-empty category* which has one, and hence all, of the following properties:

(i) *for one, and hence every, object $G \in G$, the identity functor $1_G \colon G \to G$ is naturally isomorphic to the constant functor $\mathrm{cst}_G \colon G \to G$ which sends every map of G to the identity map of the object G,*

(ii) *every object of G is an initial object, and*

(iii) *every object of G is a terminal object.*

Also not difficult to show is that

(iv) *if G is a categorically contractible category, then so is its opposite G^{op},*

(v) *if G is a categorically contractible category then so is, for every category D, the diagram category G^D* (33.2),

and last but not least

(vi) *if Y is a category, then the full subcategory of Y spanned by the initial (resp. terminal) objects is either empty or categorically contractible*

which property is behind the notion of

37.3. Universal properties. In view of 37.2(vi) one can sometimes verify the categorical uniqueness (37.1) of a non-empty set of objects in a category Y by noting that these objects are initial (or terminal) objects of Y. This is for instance what happens when one makes the somewhat imprecise statement that

(i) some objects in a category X are unique up to a unique isomorphism because they have a certain *initial* (or *terminal*) **universal property**,

as what one really means by this statement is that

(ii) these objects together with some, often not explicitly mentioned, additional structure are *initial* (or *terminal*) objects in the category of all objects of X with such additional structure.

37.4. Examples. Let Y be a category and let A, B and C be objects of Y. Then, *if they exist*,

(i) *the coproducts of A and B*, i.e. the pairs consisting of an object $Q \in Y$ and an (often suppressed) diagram in Y of the form $A \to Q \leftarrow B$ which is an *initial* object in the category of all such pairs, *are canonically isomorphic*,

(ii) dually *the products of B and C*, i.e. the pairs consisting of an object $P \in Y$ and an (often suppressed) diagram in Y of the form $B \leftarrow P \to C$ which is a *terminal* object in the category of all such pairs, *are canonically isomorphic*, and

(iii) somewhat less obviously, *the coproducts of A with a product of B and C*, i.e. the quadruples consisting of a pair of objects $P, R \in Y$ and an (often suppressed) pair of diagrams in Y of the form $A \to R \leftarrow P$ and $B \leftarrow P \to C$ such that (iii)′ the pair consisting of P and $B \leftarrow P \rightarrow C$ is a product of B and C (ii), and

(iii)″ the pair consisting of R and $A \rightarrow R \leftarrow P$ is a coproduct of A and P,

are canonically isomorphic.

In this last case the quadruples satisfying (iii)′ and (iii)″ are *initial* objects in the category of the quadruples satisfying (iii)″, but *not* in the category of all quadruples.

The obvious homotopical analog of categorical uniqueness (37.1) is the following notion of

37.5. Homotopical uniqueness. Given a homotopical category Y and a non-empty set I, we will say that objects $Y_i \in Y$ ($i \in I$) are **homotopically unique** or **canonically weakly equivalent** if the **homotopically full** subcategory of Y spanned by these objects, i.e. the full subcategory spanned by these objects and all weakly equivalent ones is *homotopically contractible* in the following sense.

37.6. Homotopically contractible categories. A homotopical category G (33.1) will be called **homotopically contractible** if

a) the unique functor $G \rightarrow [0]$ (37.2) is a *homotopical equivalence of homotopical categories* (33.1)

or equivalently

b) there exists an object $G \in G$ with the property that the identity functor $1_G \colon G \rightarrow G$ is *naturally weakly equivalent* (33.1) to the constant functor $\mathrm{cst}_G \colon G \rightarrow G$ (37.2).

This is also equivalent to the requirement that

c) G is a *non-empty* category with the property that, for *every* object $G \in G$, the identity functor $1_G \colon G \rightarrow G$ is *naturally weakly equivalent* to the constant functor $\mathrm{cst}_G \colon G \rightarrow G$

in view of the fact that iterated application of the two out of three property yields that

(i) *if G is homotopically contractible then every map in G is a weak equivalence.*

One also readily verifies that

(ii) *if G is homotopically contractible, then the homotopy category $\operatorname{Ho} C$ of C (33.8) is categorically contractible (37.2),*

(iii) *if every weak equivalence in G is an isomorphism, then G is homotopically contractible iff it is categorically contractible,* and

(iv) *if G is homotopically contractible, then so are its opposite G^{op} and, for every homotopical category D, the homotopical diagram categories G^D and $(G^D)_{\mathrm{w}}$ (33.2).*

Moreover in the next section (in 38.2) we will define notions of *homotopically initial* and *homotopically terminal* objects in a homotopical category with the properties that (as we will prove in 38.5)

(v) *if G is a homotopically contractible category, then every object of G is both homotopically initial and homotopically terminal*

and, last but not least,

> (vi) *if $\boldsymbol{Y}$ is a homotopical category, then the full subcategory of $\boldsymbol{Y}$ spanned by the homotopically initial (resp. homotopically terminal) objects is a homotopically full subcategory (37.5) which is either empty or homotopically contractible.*

This last property then suggests the following formulation of

37.7. Homotopically universal properties. In view of 37.6(vi) one can sometimes verify the homotopical uniqueness (37.5) of a non-empty set of objects in a homotopical category $\boldsymbol{Y}$ by noting that these objects are homotopically initial (or terminal) objects of $\boldsymbol{Y}$ and, as in the categorical case (37.3), it is therefore sometimes convenient to make the somewhat imprecise statement that

> (i) *some objects in a homotopical category $\boldsymbol{X}$ are homotopically unique because they have a certain homotopically initial (or terminal)* **homotopically universal property**

as a shorthand for saying that

> (ii) *these objects together with some, not necessarily explicitly mentioned, additional structure are homotopically initial (or terminal) objects in the category of all objects of $\boldsymbol{X}$ with such additional structure.*

38. Homotopically initial and terminal objects

This last section deals with the *homotopically initial* and *terminal objects* mentioned in 37.6 and 37.7 and uses them to define homotopical Kan extensions.

As it is not immediately clear how to define homotopically initial and terminal objects we start with some

38.1. Motivation. Our definition of homotopically initial and terminal objects is motivated by the following unorthodox characterization of initial and terminal objects:

> (i) *Given a category $\boldsymbol{Y}$, an object $Y \in \boldsymbol{Y}$ is initial (resp. terminal) iff there exists a natural transformation (37.2)*
>
> $$f \colon \operatorname{cst}_Y \to 1_{\boldsymbol{Y}} \qquad (\textit{resp. } f \colon 1_{\boldsymbol{Y}} \to \operatorname{cst}_Y)$$
>
> *such that the map $fY \colon Y \to Y \in \boldsymbol{Y}$ is an isomorphism*

of which one verifies the initial half by noting that every map $m \colon Y \to Y' \in \boldsymbol{Y}$ gives rise to commutative diagrams

and

of which the first implies that $fY = 1_Y$, while the second, together with this equality, yields that $m = fY'$.

This characterization suggests that, given a homotopical category $\boldsymbol{Y}$ (33.1), one considers objects $Y \in \boldsymbol{Y}$ for which there exists a zigzag of natural transformations between homotopical functors

$$\text{from } \operatorname{cst}_Y \text{ to } 1_{\boldsymbol{Y}} \qquad (\textit{resp. from } 1_{\boldsymbol{Y}} \text{ to } \operatorname{cst}_Y)$$

such that

>   (ii) the backward natural transformations are natural weak equivalences, and
>   (iii) each of the forward natural transformations sends Y to a weak equivalence in $\boldsymbol{Y}$.

Such objects clearly have the property mentioned in 37.6(v). However, in order to be able to prove that they also have the property mentioned in 37.6(vi), we had to require in addition to (ii) and (iii) that

>   (iv) all but (at most) one of the forward natural transformations are also natural weak equivalences

unless $\boldsymbol{Y}$ admits a 3-*arrow calculus* (36.1) in which case this would be a consequence of 36.3.

We therefore define as follows

38.2. Homotopically initial and terminal objects. Given a homotopical category $\boldsymbol{Y}$ (33.1), an object $Y \in \boldsymbol{Y}$ will be called **homotopically initial** (resp. **terminal**) if there exists a zigzag of natural transformations

$$\operatorname{cst}_Y \cdots F_0 \xrightarrow{f} F_1 \cdots 1_{\boldsymbol{Y}} \qquad (\text{resp. } 1_{\boldsymbol{Y}} \cdots F_1 \xrightarrow{f} F_0 \cdots \operatorname{cst}_Y)$$

in which

>   (i) the $\cdots$'s denote (possibly empty) zigzags of *natural weak equivalences*, and
>   (ii) the map $fY \in \boldsymbol{Y}$ is a *weak equivalence*.

This definition then readily implies

38.3. Proposition. *Let $\boldsymbol{Y}$ be a homotopical category and let $Y \in \boldsymbol{Y}$ be an object. Then*

>   (i) *if Y is homotopically initial (resp. terminal), then so is every object of $\boldsymbol{Y}$ which is weakly equivalent (33.2) to Y,*
>   (ii) *if Y is homotopically initial (resp. terminal), then the object $\gamma Y \in \operatorname{Ho} \boldsymbol{Y}$ (33.9) is initial (resp. terminal),*
>   (iii) *if Y is initial (resp. terminal), then Y is homotopically initial (resp. terminal), and*
>   (iv) *if every weak equivalence in $\boldsymbol{Y}$ is an isomorphism and Y is homotopically initial (resp. terminal), then Y is initial (resp. terminal).*

One also has the sometimes useful

38.4. Proposition. *Given a homotopical category $\boldsymbol{Y}$, two subcategories $\boldsymbol{Y}_1$ and $\boldsymbol{Y}_2 \subset \boldsymbol{Y}$ of which $\boldsymbol{Y}_1$ (resp. $\boldsymbol{Y}_2$) is categorically contractible (37.2) and two objects $Y_1 \in \boldsymbol{Y}_1$ and $Y_2 \in \boldsymbol{Y}_2$,*

>   (i) *a map $Y_1 \to Y_2 \in \boldsymbol{Y}$ is a homotopically initial (resp. terminal) object (38.2) of the comma category $(\boldsymbol{Y}_1 \downarrow \boldsymbol{Y}_2)$ iff*
>   (ii) *it is a homotopically initial (resp. terminal) object of its subcategory $(Y_1 \downarrow \boldsymbol{Y}_2)$ (resp. $(\boldsymbol{Y}_1 \downarrow Y_2)$).*

Proof (of the initial case). This follows readily from the observation that, for every object $Y_1' \to Y_2 \in (\mathbf{Y}_1 \downarrow \mathbf{Y}_2)$ there is a *unique* map $Y_1 \to Y_2 \in (Y_1 \downarrow \mathbf{Y}_2)$ and a *unique isomorphism* $Y_1' \approx Y_1$ such that the following diagram commutes

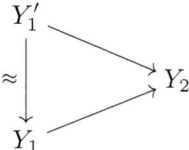

It thus remains to give a

38.5. Proof of 37.6(v) and (vi). Proposition 37.6(v) follows readily from 37.6(i) and in order to prove (the initial half of) 37.6(vi) it suffices, in view of 38.3(i), to show that, for every two homotopically initial objects $Y, Y' \in \mathbf{Y}$ and natural transformations

$$f\colon F_0 \to F_1 \quad \text{and} \quad f'\colon F_0' \to F_1'$$

as in 38.2, the map $fY' \in \mathbf{Y}$ is a weak equivalence. To do this consider the commutative diagram in $\mathbf{Y}$

$$\begin{array}{ccccccc}
F_0 F_0' F_0 Y' & \xrightarrow{fF_0' F_0 Y'} & F_1 F_0' F_0 Y' & \xrightarrow[\sim]{F_1 F_0' fY'} & F_1 F_0' F_1 Y' \\
{\scriptstyle F_0 f' F_0 Y'}\downarrow\scriptstyle\sim & & {\scriptstyle F_1 f' F_0 Y'}\downarrow & & {\scriptstyle F_1 f' F_1 Y'}\downarrow\scriptstyle\sim \\
F_0 F_1' F_0 Y' & \xrightarrow[\sim]{fF_1' F_0 Y'} & F_1 F_1' F_0 Y' & \xrightarrow{F_1 F_1' fY'} & F_1 F_1' F_1 Y'
\end{array}$$

A straightforward calculation which repeatedly uses the two out of three property (33.1) then yields that

(i) as F_0 is naturally weakly equivalent to cst_Y it sends every map and in particular the map $f'F_0 Y'$ to a weak equivalence

(ii) as F_0 is naturally weakly equivalent to cst_Y and F_1' is naturally weakly equivalent to $1_{\mathbf{Y}}$, $F_1' F_0 Y'$ is weakly equivalent to Y and hence $fF_1' F_0 Y'$ is a weak equivalence

(iii) as F_0' is naturally weakly equivalent to $\mathrm{cst}_{Y'}$, $F_0' fY'$ and therefore $F_1 F_0' fY'$ are weak equivalences,

(iv) as F_1 is naturally weakly equivalent to $1_{\mathbf{Y}}$, $F_1 Y'$ is weakly equivalent to Y' and hence $f'F_1 Y'$ and therefore $F_1 f'F_1 Y'$ are weak equivalences.

Thus the maps indicated $\sim$ are weak equivalences and the desired result now follows from the observations that

(v) in view of the *two out of six property* (33.1), the map $F_1 F_1' fY'$ is a weak equivalence

and that

(vi) as F_1 and F_1' are both naturally weakly equivalent to $1_{\mathbf{Y}}$, the map fY' is a weak equivalence iff the map $F_1 F_1' fY'$ is so.

As an application of homotopically initial and terminal objects, which, except for a brief mention in 39.3, 41.1 and 41.5, we will only need in the second half of chapter VIII (§50 and §51), we end this section with a definition of

38.6. Homotopical Kan extensions. Given two (not necessarily homotopical) functors
$$p\colon \boldsymbol{R} \longrightarrow \boldsymbol{P} \quad\text{and}\quad q\colon \boldsymbol{R} \longrightarrow \boldsymbol{Q}$$
between homotopical categories, denote by
$$(-p \downarrow q) \quad (\text{resp. } (q \downarrow -p))$$
the homotopical category which has as objects the pairs (r, e) consisting of a (not necessarily homotopical) functor $r\colon \boldsymbol{P} \to \boldsymbol{Q}$ and a natural transformation
$$e\colon rp \longrightarrow q \quad (\text{resp. } e\colon q \longrightarrow rp)$$
and which has as maps and weak equivalences $t\colon (r_1, e_1) \to (r_2, e_2)$ the natural transformations and natural weak equivalences $t\colon r_1 \to r_2$ such that $e_2(tp) = e_1$ (resp. $(tp)e_1 = e_2$), and let
$$(-p \downarrow q)_\mathrm{w} \subset (-p \downarrow q) \quad (\text{resp. } (q \downarrow -p)_\mathrm{w} \subset (q \downarrow -p))$$
denote the full subcategory spanned by the objects (r, e) for which r is a homotopical functor. The **Kan extensions** of q along p then are defined as

(i) the terminal objects of $(-p \downarrow q)$, and
(ii) the initial objects of $(q \downarrow -p)$.

The first of these, the *terminal* objects of $(-p \downarrow q)$ are the objects of $(-p \downarrow q)$ which are "as far as possible to the right" and some authors therefore refer to these Kan extensions as *right* Kan extensions (with the awkward consequence that left derived functors and left adjoints become right Kan extensions), while other authors, noting that these terminal objects of $(-p \downarrow q)$ are "closest to q from the left" call them *left* Kan extensions. To avoid this confusion we will therefore use neither terminology, but will refer to a Kan extension (r, e) as in (i) as a **terminal Kan extension** of q along p with e as its **counit** and to a Kan extension (r, e) as in (ii) as an **initial Kan extension** of q along p with e as its **unit**.

Of course the same terminological problem comes up when dealing with homotopical Kan extensions and we therefore define

(i)′ a **homotopically terminal Kan extension** of q along p as a *homotopically terminal* object (38.2) of $(-p \downarrow q)_\mathrm{w}$, and
(ii)′ a **homotopically initial Kan extension** of q along p as a *homotopically initial* object of $(q \downarrow -p)_\mathrm{w}$.

and note that (37.6(vi))

(iii) *such homotopically terminal (resp. initial) Kan extensions of q along p, if they exist, are homotopically unique* (37.5).

CHAPTER VII

Deformable Functors and Their Approximations

39. Introduction

39.1. Summary. In dealing with homotopical categories one often runs into functors between such categories which are *not* homotopical, but which still have "homotopical meaning" because they are homotopical on a so-called **left** (or **right**) **deformation retract** of the domain category. Moreover such functors, which we will call **left** (or **right**) **deformable**, frequently are part of **deformable adjunctions** i.e. adjunctions of which the left adjoint is left deformable and the right adjoint is right deformable. Our main aim in this chapter then is to

(i) show that such a left (or right) deformable functor $f\colon X \to Y$ has **left** (or **right**) **approximations**, i.e. pairs (k, a) consisting of a homotopical functor $k\colon X \to Y$ and a natural transformation
$$a\colon k \longrightarrow f \qquad (\text{or } a\colon f \longrightarrow k)$$
which are *homotopically terminal* (or *initial*) objects (38.2) in the homotopical category of all such pairs,

(ii) show that, for such a deformable adjunction $f\colon X \leftrightarrow Y : f'$, its adjunction induces, for every left approximation (k, a) of f and right approximation (k', a') of f', a **derived adjunction** (33.9)
$$\operatorname{Ho} k\colon \operatorname{Ho} X \longleftrightarrow \operatorname{Ho} Y : \operatorname{Ho} k'$$
which is natural in (k, a) and (k', a'), and

(iii) investigate the behavior of these approximations and derived adjunctions under *composition*.

In more detail:

39.2. Deformable functors and deformable adjunctions. Given a homotopical category X,

(i) a **left** (or **right**) **deformation** of X will be a pair (r, s) which consists of a *homotopical functor* $r\colon X \to X$ and a *natural weak equivalence*
$$s\colon r \to 1_X \qquad (\text{or } s\colon 1_X \to r)$$
and a **left** (or **right**) **deformation retract** of X will be a *full* subcategory $X_0 \subset X$ for which there exists a left (or right) deformation (r, s) such that X_0 contains the image of r.

We then

(ii) call a not necessarily homotopical functor $f\colon X \to Y$ between homotopical categories **left** (or **right**) **deformable** if f is *homotopical* on a *left* (or *right*) *deformation retract* of X, and

(iii) call an adjunction $f\colon \boldsymbol{X} \leftrightarrow \boldsymbol{Y} : f'$ between homotopical categories a **deformable adjunction** if the left adjoint f is left deformable and the right adjoint f' is right deformable.

A useful property of deformable functors is that they have

39.3. Approximations. A **left** (or **right**) **approximation** of a not necessarily homotopical functor $f\colon \boldsymbol{X} \to \boldsymbol{Y}$ between homotopical categories will be a homotopically terminal (or initial) Kan extension of f along $1_{\boldsymbol{X}}$ (38.6), i.e. a pair (k, a) consisting of a *homotopical* functor $k\colon \boldsymbol{X} \to \boldsymbol{Y}$ and a natural transformation

$$a\colon k \to f \qquad (\text{or } a\colon f \to k)$$

which is a homotopically terminal (or initial) object (38.2) in the homotopical category of all such pairs. Thus (37.6(vi))

(i) all such left (or right) approximations of f, they they exist, are *canonically weakly equivalent* (37.5).

Moreover

(ii) if (r, s) is a left (or right) deformation of $\boldsymbol{X}$ such that f is homotopical on the full subcategory spanned by the image of r, then the pair (fr, fs) is a left (or right) approximation of f, and hence

(iii) a *sufficient* condition for the existence of left (or right) approximations of f is that f be *left* (or *right*) *deformable* (39.2).

To next deal with the question when, for two composable deformable functors, their composition is also deformable and the compositions of their approximations are approximations of their composition we introduce the notions of

39.4. Deformable and locally deformable composable pairs of functors. Given two functors $f_1\colon \boldsymbol{X} \to \boldsymbol{Y}$ and $f_2\colon \boldsymbol{Y} \to \boldsymbol{Z}$ between homotopical categories, we call the pair (f_1, f_2) **locally left** (resp. **right**) **deformable** if there exist left (or right) deformation retracts $\boldsymbol{X}_0 \subset \boldsymbol{X}$ and $\boldsymbol{Y}_0 \subset \boldsymbol{Y}$ (39.2) such that

(i) f_1 is homotopical on $\boldsymbol{X}_0$,
(ii) f_2 is homotopical on $\boldsymbol{Y}_0$, and
(iii) $f_2 f_1$ is homotopical on $\boldsymbol{X}_0$

and call the pair (f_1, f_2) **left** (or **right**) **deformable** if there exist such deformation retracts for which in addition

(iv) f_1 sends all of $\boldsymbol{X}_0$ into $\boldsymbol{Y}_0$

(which last condition together with (i) and (ii) implies (iii)).

Now we can formulate the desired results on

39.5. Compositions of deformable functors. Given a composable pair of left (or right) deformable functors $f_1\colon \boldsymbol{X} \to \boldsymbol{Y}$ and $f_2\colon \boldsymbol{Y} \to \boldsymbol{Z}$, a *sufficient* condition in order that

(i) their composition $f_2 f_1 \colon \boldsymbol{X} \to \boldsymbol{Z}$ is also left (or right) deformable, and
(ii) all compositions of a left (or right) approximation of f_1 with a similar approximation of f_2 is such an approximation of $f_2 f_1$

is that

(iii) the pair (f_1, f_2) be *left* (or *right*) *deformable* (39.4).

However not infrequently it suffices to verify only that the pair (f_1, f_2) is deformable (39.4), because the categories and functors involved have additional properties, which allow one to apply the following result on

39.6. Compositions of deformable adjunctions. Given two composable deformable adjunctions (39.2)

$$f_1 \colon \boldsymbol{X} \longleftrightarrow \boldsymbol{Y} : f_1' \qquad \text{and} \qquad f_2 \colon \boldsymbol{X} \longleftrightarrow \boldsymbol{Y} : f_2'$$

for which the pairs $(f_1 f_2)$ and (f_2', f_1') are locally left and right deformable (39.4) respectively, a *sufficient* condition in order that

  (i) the pairs (f_1, f_2) and (f_2', f_1') are left and right deformable (39.4) respectively

is that either

  (ii) the pair (f_2', f_1') is right deformable and the homotopical category $\boldsymbol{Z}$ is *saturated* (33.9)

or dually that

  (iii) the pair (f_1, f_2) is left deformable and the homotopical category $\boldsymbol{X}$ is *saturated*.

The proof of this involves

39.7. Derived adjunctions. Given a deformable adjunctions $f \colon \boldsymbol{X} \leftrightarrow \boldsymbol{Y} : f'$ (39.2),

  (i) its adjunction induces, for every left approximation (k, a) of f and right approximation (k', a') of f' (39.3), a *derived adjunction* (33.9)

$$\operatorname{Ho} k \colon \operatorname{Ho} \boldsymbol{X} \longleftrightarrow \operatorname{Ho} \boldsymbol{Y} : \operatorname{Ho} k'$$

which is *natural* with respect to these approximations,

and given two composable deformable adjunctions

$$f_1 \colon \boldsymbol{X} \longleftrightarrow \boldsymbol{Y} : f_1' \qquad \text{and} \qquad f_2 \colon \boldsymbol{Y} \longleftrightarrow \boldsymbol{Z} : f_2'$$

for which the pairs (f_1, f_2) and (f_2', f_1') are respectively locally left and right deformable (39.4)

  (ii) the compositions of their derived adjunctions and the derived adjunctions of their composition are connected by a *conjugate* pair of natural transformations.

The second of these results then is used to prove 39.6, while we use the first in a brief discussion of

39.8. A Quillen condition. We end the chapter by showing that, given a deformable adjunction $f \colon \boldsymbol{X} \leftrightarrow \boldsymbol{Y} : f'$ (39.2),

  (i) a *sufficient* condition in order that the left approximations of f and the right approximations of f' (39.3) are homotopically inverse homotopical equivalences of homotopical categories (33.1)

is that

  (ii) the adjunction satisfies the **Quillen condition** that there exist a left deformation retract $\boldsymbol{X}_0 \subset \boldsymbol{X}$ on which f is homotopical (39.2) and a right deformation retract $\boldsymbol{Y}_0 \subset \boldsymbol{Y}$ on which f' is homotopical such that, for every pair of objects $X_0 \in \boldsymbol{X}_0$ and $Y_0 \in \boldsymbol{Y}_0$, a map $fX_0 \to Y_0 \in \boldsymbol{Y}$ is a weak equivalence iff its adjunct $X_0 \to f'Y_0 \in \boldsymbol{X}$ is so

and that

(iii) if the homotopical categories $\boldsymbol{X}$ and $\boldsymbol{Y}$ are saturated (33.9), then this condition is also *necessary*.

39.9. Organization of the chapter. After recalling, in the remainder of this section, some results on *adjoint functors* which we will need in the last three sections of this chapter as well as in the next chapter, we introduce *deformable functors* (in §40) and their *approximations* (in §41) and discuss their *compositions* (in §42). The next two sections (§43 and §44) are devoted to the *derived adjunctions* which we mention in 39.7 and the last section deals with the *Quillen condition* (39.8).

It thus remains to review some results involving

39.10. Adjunctions [Mac71, Ch. IV]. The **unit** and **counit** of an adjunction $f\colon \boldsymbol{X} \leftrightarrow \boldsymbol{Y} : f'$ are the natural transformations

$$\eta\colon 1_{\boldsymbol{X}} \longrightarrow f'f \qquad \text{and} \qquad \varepsilon\colon ff' \longrightarrow 1_{\boldsymbol{Y}}$$

which assign to an object $X \in \boldsymbol{X}$ and $Y \in \boldsymbol{Y}$ the adjuncts of the identity maps of the objects $fX \in \boldsymbol{Y}$ and $f'Y \in \boldsymbol{X}$ respectively. Some of their properties are:

(i) *the compositions*

$$f \xrightarrow{f\eta} ff'f \xrightarrow{\varepsilon f} f \qquad \text{and} \qquad f' \xrightarrow{\eta f'} f'ff' \xrightarrow{f'\varepsilon} f'$$

are identity natural transformations of f and f' respectively,

(ii) *η and ε are both natural isomorphisms iff one (and hence both) of the functors f and f' are equivalences of categories,*

(iii) *given two adjunctions*

$$f_1\colon \boldsymbol{X} \longleftrightarrow \boldsymbol{Y} : f'_1 \qquad \text{and} \qquad f_2\colon \boldsymbol{Y} \longleftrightarrow \boldsymbol{Z} : f'_2$$

with units η_1 and η_2 and counits ε_1 and ε_2, the unit and the counit of their composition $f_2f_1\colon \boldsymbol{X} \leftrightarrow \boldsymbol{Z} : f'_1f'_2$ are the compositions

$$1_{\boldsymbol{X}} \xrightarrow{\eta_1} f'_1f_1 \xrightarrow{\eta_2} f'_1f'_2f_2f_1 \qquad \text{and} \qquad f_2f_1f'_1f'_2 \xrightarrow{\varepsilon_1} f_2f'_2 \xrightarrow{\varepsilon_2} 1_{\boldsymbol{Z}} \ ,$$

and

(iv) *given two functors*

$$f\colon \boldsymbol{X} \longrightarrow \boldsymbol{Y} \qquad \text{and} \qquad f'\colon \boldsymbol{Y} \longrightarrow \boldsymbol{X}$$

and a natural transformation

$$e\colon 1_{\boldsymbol{X}} \longrightarrow f'f \qquad (\text{resp. } e\colon ff' \longrightarrow 1_{\boldsymbol{Y}}),$$

there is at most one adjunction $f\colon \boldsymbol{X} \leftrightarrow \boldsymbol{Y} : f'$ with e as its unit (resp. counit).

Given two adjunctions

$$f\colon \boldsymbol{X} \longleftrightarrow \boldsymbol{Y} : f' \qquad \text{and} \qquad g\colon \boldsymbol{X} \longleftrightarrow \boldsymbol{Y} : g'$$

two natural transformations

$$h\colon f \longrightarrow g \qquad \text{and} \qquad h'\colon g' \longrightarrow f'$$

are called **conjugate** (with respect to the given adjunctions) if, for every pair of objects $X \in \mathbf{X}$ and $Y \in \mathbf{Y}$, the following diagram

$$\begin{array}{ccc} \mathbf{Y}(gX,Y) & \approx & \mathbf{X}(X,g'Y) \\ \downarrow h^* & & \downarrow h'_* \\ \mathbf{Y}(fX,Y) & \approx & \mathbf{X}(X,f'Y) \end{array}$$

in which the horizontal isomorphisms are the adjunction isomorphisms, commutes, which is the case iff any one of the following four diagrams, in which ε_1 and ε_2 and η_1 and η_2 denote the counits and the units of these adjunctions, commutes

$$\begin{array}{ccc} f & \xrightarrow{h} & g \\ \eta_2 \downarrow & & \uparrow \varepsilon_1 \\ fg'g & \xrightarrow{h'} & ff'g \end{array} \qquad \begin{array}{ccc} g' & \xrightarrow{h'} & f' \\ \eta_1 \downarrow & & \uparrow \varepsilon_2 \\ f'fg' & \xrightarrow{h} & f'gg' \end{array}$$

$$\begin{array}{ccc} fg' & \xrightarrow{h'} & ff' \\ h \downarrow & & \downarrow \varepsilon_1 \\ gg' & \xrightarrow{\varepsilon_2} & 1_\mathbf{Y} \end{array} \qquad \begin{array}{ccc} 1_\mathbf{X} & \xrightarrow{\eta_1} & f'f \\ \eta_2 \downarrow & & \downarrow h \\ g'g & \xrightarrow{h'} & f'g \end{array}$$

Some of their properties are:

(v) *given two adjunctions*

$$f\colon \mathbf{X} \longleftrightarrow \mathbf{Y} : f' \qquad \text{and} \qquad g\colon \mathbf{X} \longleftrightarrow \mathbf{Y} : g'$$

every natural transformation $f \to g$ (resp. $g' \to f'$) has a unique conjugate and the conjugate of a natural isomorphism is also a natural isomorphism, and

(vi) *given an adjunction $f\colon \mathbf{X} \leftrightarrow \mathbf{Y} : f'$, a pair of functors*

$$g\colon \mathbf{X} \longrightarrow \mathbf{Y} \qquad \text{and} \qquad g'\colon \mathbf{Y} \longrightarrow \mathbf{X}$$

and a pair of natural isomorphisms

$$h\colon f \longrightarrow g \qquad \text{and} \qquad h'\colon g' \longrightarrow f' ,$$

there is a unique adjunction $g\colon \mathbf{X} \leftrightarrow \mathbf{Y} : g'$ such that h and h' are conjugate with respect to the adjunctions

$$f\colon \mathbf{X} \longleftrightarrow \mathbf{Y} : f' \qquad \text{and} \qquad g\colon \mathbf{X} \longleftrightarrow \mathbf{Y} : g' .$$

40. Deformable functors

In this section we introduce *left* and *right deformable functors* and discuss some immediate consequences of their definition.

We start with the auxiliary notions of

40.1. Left and right deformations and deformation retracts. A **left** (resp. a **right**) **deformation** of a homotopical category $\mathbf{X}$ (33.1) will be a pair (r, s) consisting of

(i) a *homotopical* functor $r\colon \mathbf{X} \to \mathbf{X}$, and
(ii) a *natural weak equivalence*

$$s\colon r \to 1_{\mathbf{X}} \qquad (\text{resp. } s\colon 1_{\mathbf{X}} \to r)$$

and a **left** (resp. a **right**) **deformation retract** of $\mathbf{X}$ will be a *full* subcategory $\mathbf{X}_0 \subset \mathbf{X}$ for which there exists a left (resp. a right) deformation (r, s) of $\mathbf{X}$ **into** $\mathbf{X}_0$, i.e. such that $fX \in \mathbf{X}_0$ for every object $X \in \mathbf{X}$.

An immediate consequence of this definition is that

(iii) *if $\mathbf{X}_0$ is a left or right deformation retract of $\mathbf{X}$, then so is every full subcategory of $\mathbf{X}$ which contains $\mathbf{X}_0$.*

Now we can define

40.2. Left and right deformable functors. A (not necessarily homotopical) functor $f\colon \mathbf{X} \to \mathbf{Y}$ between homotopical categories will be called **left** (resp. **right**) **deformable** if

(i) there exists a **left** (resp. a **right**) **f-deformation retract**, i.e. a left (resp. a right) deformation retract $\mathbf{X}_0 \subset \mathbf{X}$ (40.1) on which f is homotopical.

It is sometimes convenient to have a somewhat more explicit description of such deformability and we therefore also consider the notion of a **left** (resp. a **right**) **f-deformation**, i.e. a left (resp. a right) deformation (r, s) of $\mathbf{X}$ (40.1) such that

(ii) *f is homotopical on the full subcategory spanned by the image of the functor r*

or equivalently

(ii)′ *the functor fr is a homotopical functor and the natural transformation fsr is a natural weak equivalence,*

so that

(iii) *a functor $f\colon \mathbf{X} \to \mathbf{Y}$ is left (resp. right) deformable iff there exists a left (resp. a right) f-deformation.*

Moreover clearly

(iv) *every homotopical functor is both left and right deformable, and*
(v) *if $f_1, f_2\colon \mathbf{X} \to \mathbf{Y}$ are functors between homotopical categories which are naturally weakly equivalent (33.1), and one of them is left or right deformable, then so is the other.*

Next we note that following "f-deformation retract version" of 40.1(iii).

40.3. Proposition. *Let $f\colon \mathbf{X} \to \mathbf{Y}$ be a left (resp. a right) deformable functor, let $\mathbf{X}_0 \subset \mathbf{X}$ be a left (resp. a right) f-deformation retract (40.2) and let $\mathbf{X}_1 \subset \mathbf{X}$ be a full subcategory containing $\mathbf{X}_0$. Then $\mathbf{X}_1$ is a left (resp. a right) f-deformation retract iff, for every object $X_p \in \mathbf{X}_1$, there exists an object $X_{p,0} \in \mathbf{X}_0$ and a weak equivalence*

$$b_p\colon X_{p,0} \to X_p \qquad (\text{resp. } b_p\colon X_p \to X_{p,0})$$

such that the map $fb_p \in \mathbf{Y}$ is also a weak equivalence.

Proof (of the left half). The "only if" part of the proposition is obvious and to prove the "if" part one has to show that, for every pair of objects $X_p, X_q \in \boldsymbol{X}_1$ and weak equivalences $a\colon X_p \to X_q \in \boldsymbol{X}$, the map $fa \in \boldsymbol{Y}$ is also a weak equivalence. To do this one chooses a left deformation (r,s) of $\boldsymbol{X}$ into $\boldsymbol{X}_0$ and notes that all the other outside maps in the commutative diagram

$$\begin{array}{ccccccc}
frX_{p,0} & \xrightarrow{frb_p} & frX_p & \xrightarrow{fra} & frX_q & \xleftarrow{frb_q} & frX_{q,0} \\
{\scriptstyle fsX_{p,0}}\downarrow & & {\scriptstyle fsX_p}\downarrow & & {\scriptstyle fsX_q}\downarrow & & {\scriptstyle fsX_{q,0}}\downarrow \\
fX_{p,0} & \xrightarrow{fb_p} & fX_p & \xrightarrow{fa} & fX_q & \xleftarrow{fb_q} & fX_{q,0}
\end{array}$$

are weak equivalences and applies the two out of three property (33.1).

Somewhat less obvious is the

40.4. Existence of a unique maximal f-deformation retract. *Let $f\colon \boldsymbol{X} \to \boldsymbol{Y}$ be a left (resp. a right) f-deformable functor. Then*

(i) *the full subcategory of $\boldsymbol{X}$ spanned by the left (resp. the right) f-deformation retracts is itself a left (resp. a right) f-deformation retract, and*

(ii) *if $\boldsymbol{X}_0 \subset \boldsymbol{X}$ is any left (resp. right) f-deformation retract then this unique maximal left (resp. right) f-deformation retract (see (i)) is the full subcategory of $\boldsymbol{X}$ spanned by the objects $X_p \in \boldsymbol{X}$ for which there exists an object $X_{p,0} \in \boldsymbol{X}_0$ and a weak equivalence*

$$b_p \colon X_{p,0} \to X_p \qquad (\text{resp. } b_p \colon X_p \to X_{p,0})$$

such that the map $fb_p \in \boldsymbol{Y}$ is also a weak equivalence.

Proof (of the left half). Part (ii) is an immediate consequence of 40.3 and part (i) and to prove (i) it suffices to show that, for every two left f-deformation retracts $\boldsymbol{X}_1$ and $\boldsymbol{X}_2$ and every weak equivalence $w\colon X_1 \to X_2 \in \boldsymbol{X}$ with $X_1 \in \boldsymbol{X}_1$ and $X_2 \in \boldsymbol{X}_2$, the map $fw\colon fX_1 \to fX_2 \in \boldsymbol{Y}$ is also a weak equivalence. To do this one chooses left f-deformations (r_1, s_1) and (r_2, s_2) of $\boldsymbol{X}$ into $\boldsymbol{X}_1$ and $\boldsymbol{X}_2$ respectively (40.1) and notes that in the commutative diagram in $\boldsymbol{Y}$

$$\begin{array}{ccccc}
fr_1 r_2 X_1 & \xrightarrow{fs_1 r_2 X_1} & fr_2 X_1 & \xrightarrow[\sim]{fr_2 w} & fr_2 X_2 \\
{\scriptstyle fr_1 s_2 X_1}\downarrow{\scriptstyle \sim} & & {\scriptstyle fs_2 X_1}\downarrow & & {\scriptstyle fs_2 X_2}\downarrow{\scriptstyle \sim} \\
fr_1 X_1 & \xrightarrow[\sim]{fs_1 X_1} & fX_1 & \xrightarrow{fw} & fX_2
\end{array}$$

the maps indicated $\sim$ are weak equivalences. The desired result then is an immediate consequence of the *two out of six property* (33.1).

A somewhat similar result for f-deformations is

40.5. Homotopical uniqueness of f-deformations. *Let $f\colon \boldsymbol{X} \to \boldsymbol{Y}$ be a left (resp. a right) deformable functor. Then the homotopical category which has*

(i) *as objects the left (resp. the right) deformations of $\boldsymbol{X}$ into $\boldsymbol{X}_0$, and*

(ii) *for every two such deformations (r_1, s_1) and (r_2, s_2), as maps and weak equivalences $(r_1, s_1) \to (r_2, s_2)$ the natural transformations and natural weak equivalences $t\colon r_1 \to r_2$ such that $s_2 t = s_1$ (resp. $t s_1 = s_2$)*

is homotopically contractible (37.6) *and so is, for every left (resp. right) f-deformation retract $X_0 \subset X$* (40.2), *its full subcategory spanned by the left (resp. right) f-deformations of X into X_0.*

Proof (of the left half). This follows readily from the observation that every two left deformations (r_1, s_1) and (r_2, s_2) give rise to a zigzag

$$(r_1, s_1) \xleftarrow{r_1 s_2} (r_1 r_2, s_1 s_2) \xrightarrow{s_1 r_2} (r_2, s_2)$$

in which $s_1 s_2$ denotes the diagonal of the commutative diagram

$$\begin{array}{ccc} r_1 r_2 & \xrightarrow{s_1} & r_2 \\ {\scriptstyle s_2}\downarrow & & \downarrow{\scriptstyle s_2} \\ r_1 & \xrightarrow{s_1} & 1 \end{array}$$

which is natural in both (r_1, s_1) and (r_2, s_2).

We end with a brief mention of the notion of

40.6. Left and right deformable natural transformations. Given two homotopical categories X and Y, a natural transformation $h\colon f \to g$ between (not necessarily homotopical) functors $X \to Y$ will be called **left** (resp. **right**) **deformable** if there exists a **left** (resp. a **right**) h-**deformation retract**, i.e. a left (resp. a right) deformation retract $X_0 \subset X$ on which both f and g are homotopical, and we similarly will call a left (resp. a right) deformation (r, s) of X a **left** (resp. a **right**) h-**deformation** if (r, s) is both a left (resp. a right) f-deformation and g-deformation.

Note that, if h is a *natural weak equivalence*, then the following three statements are equivalent:

(i) (r, s) is a left (resp. a right) f-deformation,
(ii) (r, s) is a left (resp. a right) g-deformation and
(iii) (r, s) is a left (resp. a right) h-deformation.

41. Approximations

Next we consider, for a (not necessarily homotopical) functor $f\colon X \to Y$ between homotopical categories, its *left* and *right approximations*, i.e. *homotopical functors* $X \to Y$ which, in a homotopical sense, are closest to f from the left or from the right, and note in particular that

(i) if they exist, then such left or right approximations of f are *homotopically unique* (37.5),
(ii) a *sufficient* condition for their existence is that the functor $f\colon X \to Y$ be left or right deformable (40.2), and
(iii) if f is left or right deformable, then its left or right approximations are a kind of "*not* in the homotopy category" version of Quillen's total left and right derived functors.

We thus start with defining

41.1. Left and right approximations. Let $f\colon \boldsymbol{X} \to \boldsymbol{Y}$ be a (not necessarily homotopical) functor between homotopical categories, i.e. an object of the homotopical functor category $\mathrm{Fun}(\boldsymbol{X}, \boldsymbol{Y})$ (33.2). A **left** (resp. a **right**) **approximation** of f then will be a *homotopically terminal* (resp. *initial*) *Kan extension* of f along $1_{\boldsymbol{X}}$, i.e. (38.6) a *homotopically terminal* (resp. *initial*) (38.2) object of the "homotopical category of homotopical functors $\boldsymbol{X} \to \boldsymbol{Y}$ over (resp. under) f", i.e. the homotopical category

$$\big(\mathrm{Fun}_{\mathrm{w}}(\boldsymbol{X},\boldsymbol{Y}) \downarrow f\big) \qquad \big(\text{resp. } \big(f \downarrow \mathrm{Fun}_{\mathrm{w}}(\boldsymbol{X},\boldsymbol{Y})\big)\big)$$

which has

(i) as objects the pairs (k, a) consisting of a *homotopical* functor $k\colon \boldsymbol{X} \to \boldsymbol{Y}$ (i.e. an object of the homotopical functor category $\mathrm{Fun}_{\mathrm{w}}(\boldsymbol{X}, \boldsymbol{Y})$ (33.2)) and a natural transformation

$$a\colon k \to f \qquad (\text{resp. } a\colon f \to k)$$

and

(ii) for every pair of objects (k_1, a_1) and (k_2, a_2) as maps and weak equivalences $(k_1, a_1) \to (k_2, a_2)$ the natural transformations and natural weak equivalences $t\colon k_1 \to k_2$ such that $a_2 t = a_1$ (resp. $t a_1 = a_2$).

Thus (37.5 and 37.6(vi))

(iii) *such a left (resp. right) approximation of f, if it exists, is homotopically unique (37.5).*

We now prove the promised existence of

41.2. Existence of approximations of deformable functors. *Let $f\colon \boldsymbol{X} \to \boldsymbol{Y}$ be a left (resp. a right) deformable functor (40.2). Then*

(i) *the functor f has left (resp. right) approximations.*

In fact, if (40.2) (r, s) *is a left (resp. a right) f-deformation of $\boldsymbol{X}$ then*

(ii) *the pair (fr, fs) is a left (resp. a right) approximation of f.*

Proof (of the left half). Every object $(k, a) \in (\mathrm{Fun}_{\mathrm{w}}(\boldsymbol{X}, \boldsymbol{Y}) \downarrow f)$ gives rise to a zigzag

$$(k, a) \xleftarrow{ks} (kr, as) \xrightarrow{ar} (fr, fs)$$

in which as denotes the diagonal of the commutative diagram

$$\begin{array}{ccc} kr & \xrightarrow{ar} & fr \\ {\scriptstyle ks}\downarrow & & \downarrow{\scriptstyle fs} \\ k & \xrightarrow{a} & f \end{array}$$

which is natural in (k, a) and in which

(i) the map ks is a weak equivalence, and
(ii) the map ar is so whenever $(k, a) = (fr, fs)$, in which case $ar = fsr$

and (40.2(ii)′) this readily implies (ii).

A special case is provided by

41.3. Approximations of homotopical functors. *If $f\colon \mathbf{X} \to \mathbf{Y}$ is a homotopical functor, then clearly*

(i) *the pair $(f, 1_f)$ consisting of f and its identity natural transformation is both a left and a right approximation of f,*

and more generally (37.6)

(ii) *a pair (k, a) consisting of a homotopical functor $k\colon \mathbf{X} \to \mathbf{Y}$ and a natural transformation $a\colon k \to f$ (resp. $a\colon f \to k$) is a left (resp. a right) approximation of f iff the natural transformation a is a natural weak equivalence.*

In particular

(iii) *for every homotopical category $\mathbf{X}$, the left (resp. the right) approximations of the identity functor $1_{\mathbf{X}}\colon \mathbf{X} \to \mathbf{X}$ are exactly the left (resp. the right) deformations (40.1) of $\mathbf{X}$.*

Another immediate consequence of 41.2 is

41.4. Proposition. *Let $f\colon \mathbf{X} \to \mathbf{Y}$ be a left (resp. a right) deformable functor (40.2), let $\gamma'\colon \mathbf{Y} \to \operatorname{Ho}\mathbf{Y}$ denote the localization functor of $\mathbf{Y}$ (33.9) and consider $\operatorname{Ho}\mathbf{Y}$ as a minimal homotopical category (33.6). Then*

(i) *the composition $\gamma' f\colon \mathbf{X} \to \operatorname{Ho}\mathbf{Y}$ is left (resp. right) deformable.*

In fact,

(ii) *every left (resp. right) f-deformation (40.2) is also a left (resp. a right) $\gamma' f$-deformation and hence (37.6 and 41.2)*

(iii) *for every left (resp. right) approximation (k, a) of f (41.1), the pair $(\gamma' k, \gamma' a)$ is a left (resp. a right) approximation of $\gamma' f$.*

And from this in turn we deduce the following

41.5. Relation to Quillen's total derived functors. *Let $f\colon \mathbf{X} \to \mathbf{Y}$ be a left (resp. a right) deformable functor (40.2), let $\gamma\colon \mathbf{X} \to \operatorname{Ho}\mathbf{X}$ and $\gamma'\colon \mathbf{Y} \to \operatorname{Ho}\mathbf{Y}$ be the localization functors of $\mathbf{X}$ and $\mathbf{Y}$ (33.9) and let (k, a) be a left (resp. a right) approximation of f (41.1). Then the pair $(\operatorname{Ho} k, \gamma' a)$ consisting of the functor $\operatorname{Ho} k\colon \operatorname{Ho}\mathbf{X} \to \operatorname{Ho}\mathbf{Y}$ and the natural transformation*

$$\gamma' a\colon (\operatorname{Ho} k)\gamma = \gamma' k \longrightarrow \gamma' f \qquad (\text{resp. } \gamma' a\colon \gamma' f \longrightarrow \gamma' k = (\operatorname{Ho} k)\gamma)$$

*is a **total left** (resp. **right**) **derived functor** of f in the sense of Quillen* [**Qui67**] *(or, in the terminology of 38.6, a **terminal** (resp. an **initial**) **Kan extension** of $\gamma' f$ along γ), i.e. the pair $((\operatorname{Ho} k)\gamma, \gamma' a)$ is a terminal (resp. an initial) object of the category*

$$\bigl(\operatorname{Fun}(\operatorname{Ho}\mathbf{X}, \operatorname{Ho}\mathbf{Y}) \downarrow \gamma' f\bigr) \qquad \bigl(\text{resp. } (\gamma' f \downarrow \operatorname{Fun}(\operatorname{Ho}\mathbf{X}, \operatorname{Ho}\mathbf{Y}))\bigr)$$

which has as objects the pairs (n, c) which consist of a functor $n\colon \operatorname{Ho}\mathbf{X} \to \operatorname{Ho}\mathbf{Y}$ and a natural transformation

$$c\colon n\gamma \longrightarrow \gamma' f \qquad (\text{resp. } c\colon \gamma' f \longrightarrow n\gamma)$$

and as maps $(n_1, c_1) \to (n_2, c_2)$ the natural transformations $t\colon n_1 \to n_2$ such that $c_2(t\gamma) = c_1$ (resp. $(t\gamma)c_1 = c_2$).

41. APPROXIMATIONS

Proofs of (the left half of) 41.5.
This follows from 41.4 and the observations that

(i) $\bigl(\mathrm{Fun}_w(\boldsymbol{X}, \mathrm{Ho}\,\boldsymbol{Y}) \downarrow \gamma' f\bigr)$ is a minimal homotopical category (33.6) and hence (38.3) its homotopically terminal objects are its terminal objects,

and that (33.9(ii))

(ii) composition with γ induces an isomorphism
$$\bigl(\mathrm{Fun}(\mathrm{Ho}\,\boldsymbol{X}, \mathrm{Ho}\,\boldsymbol{Y}) \downarrow \gamma' f\bigr) \approx \bigl(\mathrm{Fun}_w(\boldsymbol{X}, \mathrm{Ho}\,\boldsymbol{Y}) \downarrow \gamma' f\bigr).$$

We end with a mention of

41.6. Left and right approximations of natural transformations. Given two homotopical categories $\boldsymbol{X}$ and $\boldsymbol{Y}$ and a natural transformation $h\colon f \to g$ between (not necessarily homotopical) functors $\boldsymbol{X} \to \boldsymbol{Y}$, let
$$\bigl(\mathrm{Nat}_w(\boldsymbol{X}, \boldsymbol{Y}) \downarrow h\bigr) \qquad \bigl(\text{resp. } \bigl(h \downarrow \mathrm{Nat}_w(\boldsymbol{X}, \boldsymbol{Y})\bigr)\bigr)$$
denote "the homotopical category of the natural transformations between *homotopical* functors $\boldsymbol{X} \to \boldsymbol{Y}$ over (resp. under) h", i.e. the homotopical category which has

(i) as objects the commutative diagrams of the form

$$\begin{array}{ccc} k_1 & \xrightarrow{a_1} & f \\ {\scriptstyle k}\downarrow & & \downarrow{\scriptstyle h} \\ k_2 & \xrightarrow{a_2} & g \end{array} \qquad \left(\text{resp. } \begin{array}{ccc} f & \xrightarrow{a_1} & k_1 \\ {\scriptstyle h}\downarrow & & \downarrow{\scriptstyle k} \\ g & \xrightarrow{a_2} & k_2 \end{array}\right)$$

in which k_1 and k_2 are *homotopical* functors $\boldsymbol{X} \to \boldsymbol{Y}$ (and which we will denote by $k\colon (k_1, a_1) \to (k_2, a_2)$),

(ii) as maps $\bigl(k\colon (k_1, a_1) \to (k_2, a_2)\bigr) \to \bigl(\bar{k}\colon (\bar{k}_1, \bar{a}_1) \to (\bar{k}_2, \bar{a}_2)\bigr)$ the commutative diagrams of the form

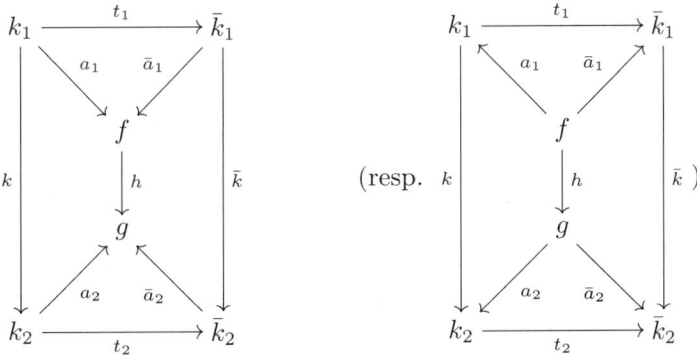

and

(iii) as weak equivalences the diagrams as in (ii) in which t_1 and t_2 are natural weak equivalences.

A **left** (resp. a **right**) **approximation** of h then will be an object $k\colon (k_1, a_1) \to (k_2, a_2)$ of
$$\bigl(\mathrm{Nat}_w(\boldsymbol{X}, \boldsymbol{Y}) \downarrow h\bigr) \qquad \bigl(\text{resp. } \bigl(h \downarrow \mathrm{Nat}_w(\boldsymbol{X}, \boldsymbol{Y})\bigr)\bigr)$$
such that

(iv) k is a homotopically terminal (resp. initial) object, and

(v) the pairs (k_1, a_1) and (k_2, a_2) are left (resp. right) approximations of f and g (41.1) respectively.

It then readily follows from 37.6(vi) and 38.3(i) that

(vi) *if one object of* $(\mathrm{Nat}_w(X, Y) \downarrow h)$ *(resp.* $(h \downarrow \mathrm{Nat}_w(X, Y)))$ *satisfies* (iv) *and* (v), *then every object that satisfies one of* (iv) *and* (v) *also satisfies the other*

so that (37.6(vi))

(vii) *such a left (resp. right) approximation of h, if it exists, is homotopically unique* (37.5).

Moreover

(viii) a *sufficient* condition for its existence is that h be *left* (resp. *right*) deformable (40.6)

as the argument used in the proof of 41.2 yields that

(ix) *for every left (resp. right) h-deformation (r, s)* (40.6)

$$hr \colon (fr, fs) \longrightarrow (gr, gs)$$

is a left (resp. a right) approximation of h.

42. Compositions

Our aim in this section is to find *sufficient conditions* on a composable pair of left (or right) deformable functors (40.2) in order that

(i) their composition is also left (or right) deformable, and

(ii) every composition of their left (or right) approximations is a left (or right) approximation of their composition.

To do this we start with defining

42.1. Compositions of approximations. Let $f_1 \colon X \to Y$ and $f_2 \colon Y \to Z$ be (not necessarily homotopical) functors between homotopical categories. The functors and natural transformations involved in the definition of approximations then give rise to *homotopical* **composition functors** (41.1)

$$\bigl(\mathrm{Fun}_w(X, Y) \downarrow f_1\bigr) \times \bigl(\mathrm{Fun}_w(Y, Z) \downarrow f_2\bigr) \longrightarrow \bigl(\mathrm{Fun}_w(X, Z) \downarrow f_2 f_1\bigr)$$

and

$$\bigl(f_1 \downarrow \mathrm{Fun}_w(X, Y)\bigr) \times \bigl(f_2 \downarrow \mathrm{Fun}_w(Y, Z)\bigr) \longrightarrow \bigl(f_2 f_1 \downarrow \mathrm{Fun}_w(X, Z)\bigr)$$

which sends an object $\bigl((k_1, a_1), (k_2, a_2)\bigr)$ to the object $(k_2 k_1, a_2 a_1)$ in which $a_2 a_1$ denotes the diagonals of

$$\begin{array}{ccc} k_2 k_1 & \xrightarrow{a_2} & f_2 k_1 \\ {\scriptstyle a_1} \downarrow & & \downarrow {\scriptstyle a_1} \\ k_2 f_1 & \xrightarrow{a_2} & f_2 f_1 \end{array} \quad \text{and} \quad \begin{array}{ccc} f_2 f_1 & \xrightarrow{a_2} & k_2 f_1 \\ {\scriptstyle a_1} \downarrow & & \downarrow {\scriptstyle a_1} \\ f_2 k_1 & \xrightarrow{a_2} & k_2 k_1 \end{array}$$

respectively.

The application of these functors to a pair of approximations does, in general, *not* produce an approximation. However one has the following

42.2. All or none proposition. *If, given a composable pair $f_1\colon X \to Y$ and $f_2\colon Y \to Z$ of (not necessarily homotopical) functors between homotopical categories, there exist a left (resp. a right) approximation of f_1 and a left (resp. a right) approximation of f_2 whose composition is a left (resp. a right) approximation of $f_2 f_1$, then every composition of a left (resp. a right) approximation of f_1 with a similar approximation of f_2 is such an approximation of $f_2 f_1$.*

Proof. This follows readily from the definition of approximations (41.1) and the fact that, in view of 37.6(i) and (vi) and 38.3(i), *every homotopical functor $C \to D$ sends either all or none of the homotopically terminal or initial objects of C to similar objects of D*.

The desired sufficient conditions involve the following notions of

42.3. Deformable and locally deformable composable pairs of functors. Given a composable pair $f_1\colon X \to Y$ and $f_2\colon Y \to Z$ of (not necessarily homotopical) functors between homotopical categories, merely assuming that the functors f_1, f_2 and $f_2 f_1$ are all three left (resp. right) deformable does not seem to be very useful and we therefore call the pair (f_1, f_2) **locally left** (resp. **right**) **deformable** if it has the *slightly stronger* property that there exist left (resp. right) deformation retract $X_0 \subset X$ and $Y_0 \subset Y$ (40.1) such that

(i) f_1 is homotopical on X_0 (and hence f_1 is left (resp. right) deformable (40.2)),

(ii) f_2 is homotopical on Y_0 (and hence f_2 is left (resp. right) deformable), and

(iii) $f_2 f_1$ is homotopical on X_0 (and hence $f_2 f_1$ is left (resp. right) deformable)

and call the pair (f_1, f_2) **left** (resp. **right**) **deformable** if it has the *even stronger* property that *in addition*

(iv) f_1 sends all of X_0 into Y_0,

which together with (i) and (ii) implies (iii).

One then readily verifies that such left (resp. right) deformability is equivalent to the requirement that

(v) there exists a left (resp. a right) f_1-deformation (r_1, s_1) (40.2) and a left (resp. a right) f_2-deformation (r_2, s_2) such that the natural transformation

$$f_2 s_2 f_1 r_1 \colon f_2 r_2 f_1 r_1 \longrightarrow f_2 f_1 r_1 \quad (\text{resp. } f_2 s_2 f_1 r_1 \colon f_2 f_1 r_1 \longrightarrow f_2 r_2 f_1 r_1)$$

is a *natural weak equivalence*,

as in that case, in view of 40.3, f_2 is homotopical on the full subcategory of Y spanned by the images of r_2 and $f_1 r_1$.

In view of 42.2 this last observation (42.3(v)) together with 41.2(ii) implies the desired

42.4. Sufficient conditions for composability of approximations. *Given (not necessarily homotopical) functors $f_1\colon X \to Y$ and $f_2\colon Y \to Z$, consider the following statements:*

  (i) *the pair (f_1, f_2) is left (resp. right) deformable (42.3),*
  (ii) *the pair (f_1, f_2) is locally left (resp. right) deformable (42.3), and*
  (iii) *every composition (42.1) of a left (resp. a right) approximation of f_1 with a similar approximation of f_2 is such an approximation of $f_2 f_1$.*

Then (i) *implies* (ii) *while conversely (in view of 37.6(i))* (ii) *and* (iii) *together imply* (i).

We end with noting that, if one wants to prove the *deformability* of a composable pair of deformable functors (i.e. the existence of deformation retracts satisfying 42.3(i)–(iv)) and these functors have the property that

  (i) they are each one of an adjoint pair of deformable functors of which the left adjoint is left deformable and the right adjoint is right deformable,
  (ii) the pair consisting of their adjoints are known to be (or easily verified to be) deformable, and
  (iii) the codomain of their composition is saturated (33.9),

then the following proposition implies that it suffices to merely verify the *local deformability* of the pair (i.e. the existence of deformation retracts satisfying 42.3(i)–(iii)).

42.5. Sufficient conditions for deformability of pairs. *Given two composable adjunctions*

$$f_1\colon X \longleftrightarrow Y : f_1' \quad \text{and} \quad f_2\colon Y \longleftrightarrow Z : f_2'$$

of functors between homotopical categories such that the pair (f_1, f_2) is locally left deformable (42.3) and the pair (f_2', f_1') is locally right deformable,

  (i) *if the category Z is saturated (33.9) and the pair (f_2', f_1') is right deformable (42.3), then the pair (f_1, f_2) is left deformable,*

and dually

  (ii) *if the category X is saturated and the pair (f_1, f_2) is left deformable, then the pair (f_2', f_1') is right deformable*

and hence

  (iii) *if the categories X and Z are both saturated then the pair (f_1, f_2) is left deformable iff the pair (f_2', f_1') is right deformable.*

Proof. In view of 33.9, 42.3(v) and 42.4 this follows readily from the following technical result.

42.6. Proposition. *Let*

$$f_1\colon X \longleftrightarrow Y : f_1' \quad \text{and} \quad f_2\colon Y \longleftrightarrow Z : f_2'$$

be adjoint pairs of functors for which the pairs (f_1, f_2) and (f_2', f_1') are locally left and right deformable (42.3) respectively and let

$$(r_1, s_1), \quad (r_2, s_2), \quad (r_1', s_1') \quad \text{and} \quad (r_2', s_2')$$

respectively be a left f_1- as well as a left f_2f_1-deformation, a left f_2-deformation, a right f_1'-deformation and a right f_2'- as well as a right $f_1'f_2'$-deformation (40.2). Then the natural transformations

$$\operatorname{Ho}(f_2 s_2 f_1 r_1)\colon \operatorname{Ho}(f_2 r_2 f_1 r_1) \longrightarrow \operatorname{Ho}(f_2 f_1 r_1) \quad \text{and}$$
$$\operatorname{Ho}(f_1' s_1' f_2' r_2')\colon \operatorname{Ho}(f_1' f_2' r_2') \longrightarrow \operatorname{Ho}(f_1' r_1' f_2' r_2')$$

have the property that, if one of them is a natural isomorphism, then so is the other.

Proof. This follows from 39.10(v) and the special case of 44.4 below considered in its proof.

43. Induced partial adjunctions

In preparation for the next section, where in 44.4 we complete the proof of 42.6, we first

(i) observe that, for every *adjoint pair of functors* between homotopical categories of which the left adjoint is left deformable and the right adjoint is right deformable, its adjunction induces a "partial" adjunction between the homotopy categories, and

(ii) discuss the behavior of these partial adjunctions under composition.

In more detail:

43.1. Deformable adjunctions. An adjunction $f\colon \boldsymbol{X} \leftrightarrow \boldsymbol{Y} : f'$ between homotopical categories will be called a **deformable adjunction** if the left adjoint f is left deformable (40.2) and the right adjoint f' is right deformable.

Similarly (40.6) given two deformable adjunctions $f\colon \boldsymbol{X} \leftrightarrow \boldsymbol{Y} :f'$ and $g\colon \boldsymbol{X} \leftrightarrow \boldsymbol{Y} :g'$, a conjugate pair of natural transformations $h\colon f \to g$ and $h'\colon g' \to f'$ between them (39.10) will be called a **deformable conjugate pair of natural transformations** if the natural transformation $h\colon f \to g$ between the left adjoints is left deformable (40.6) and the natural transformation $h'\colon g' \to f'$ between the right adjoints is right deformable.

The following result then implicitly defines

43.2. Induced partial adjunctions. *Let $\boldsymbol{X}$ and $\boldsymbol{Y}$ be homotopical categories with localization functors $\gamma\colon \boldsymbol{X} \to \operatorname{Ho}\boldsymbol{X}$ and $\gamma'\colon \boldsymbol{Y} \to \operatorname{Ho}\boldsymbol{Y}$, and let $f\colon \boldsymbol{X} \leftrightarrow \boldsymbol{Y} : f'$ be a deformable adjunction (43.1). Then*

(i) *for every left f- and right f'-deformation retract $\boldsymbol{X}_0 \subset \boldsymbol{X}$ and $\boldsymbol{Y}_0 \subset \boldsymbol{Y}$ (40.2), there exist, for every pair of objects $X \in \boldsymbol{X}$ and $Y_0 \in \boldsymbol{Y}_0$ and every pair of objects $X_0 \in \boldsymbol{X}_0$ and $Y \in \boldsymbol{Y}$, unique **partial adjunction functions***

$$\operatorname{Ho}\boldsymbol{Y}(fX, Y_0) \xrightarrow{\phi} \operatorname{Ho}\boldsymbol{X}(X, f'Y_0) \quad \text{and} \quad \operatorname{Ho}\boldsymbol{X}(X_0, f'Y) \xrightarrow{\psi} \operatorname{Ho}\boldsymbol{Y}(fX_0, Y)$$

which are natural in both variables and which are compatible with the given adjunction in the sense that the diagrams

$$\begin{array}{ccc} \boldsymbol{Y}(fX,Y_0) \xrightarrow{\approx} \boldsymbol{X}(X,f'Y_0) & & \boldsymbol{X}(X_0,f'Y) \xrightarrow{\approx} \boldsymbol{Y}(fX_0,Y) \\ \gamma' \downarrow \quad\quad\quad \downarrow \gamma & \text{and} & \gamma \downarrow \quad\quad\quad \downarrow \gamma' \\ \operatorname{Ho}\boldsymbol{Y}(fX,Y_0) \xrightarrow{\phi} \operatorname{Ho}\boldsymbol{X}(X,f'Y_0) & & \operatorname{Ho}\boldsymbol{X}(X_0,f'Y) \xrightarrow{\psi} \operatorname{Ho}\boldsymbol{Y}(fX_0,Y) \end{array}$$

in which the top maps are the adjunction isomorphisms of the given adjunction, commute,

This terminology is justified by the fact that

(ii) *for every pair of objects $X_0 \in \boldsymbol{X}_0$ and $Y_0 \in \boldsymbol{Y}_0$, the functions*

$$\operatorname{Ho}\boldsymbol{Y}(fX_0,Y_0) \xrightarrow{\phi} \operatorname{Ho}\boldsymbol{X}(X_0,f'Y_0) \quad \text{and} \quad \operatorname{Ho}\boldsymbol{X}(X_0,f'Y_0) \xrightarrow{\psi} \operatorname{Ho}\boldsymbol{Y}(fX_0,Y_0)$$

are inverse **partial adjunction isomorphisms**

while the uniqueness of these partial adjunction functions associated with $\boldsymbol{X}_0$ and $\boldsymbol{Y}_0$ implies that

(iii) *these functions are the restriction to $\boldsymbol{X} \times \boldsymbol{Y}_0$ and $\boldsymbol{X}_0 \times \boldsymbol{Y}$ of the partial adjunction functions associated with the unique (40.4) maximal left f- and right f'-deformation retracts $\boldsymbol{X}_f$ and $\boldsymbol{Y}_{f'}$.*

Moreover these partial adjunctions are also natural in the given adjunction in the sense that

(iv) *if $f\colon \boldsymbol{X} \leftrightarrow \boldsymbol{Y} :f'$ and $g\colon \boldsymbol{X} \leftrightarrow \boldsymbol{Y} :g'$ are deformable adjunctions, $h\colon f \to g$ and $h'\colon g' \to f'$ are a deformable conjugate pair of natural transformations between them (43.1) and $\boldsymbol{X}_h \subset \boldsymbol{X}$ and $\boldsymbol{Y}_{h'} \subset \boldsymbol{Y}$ denote left h and right h'-deformation retracts (40.6), then, for every pair of objects $X \in \boldsymbol{X}$ and $Y_0 \in \boldsymbol{Y}_{h'}$ and every pair of objects $X_0 \in \boldsymbol{X}_h$ and $Y \in \boldsymbol{Y}$, the diagrams*

$$\begin{array}{ccc} \operatorname{Ho}\boldsymbol{Y}(gX,Y_0) \xrightarrow{\phi} \operatorname{Ho}\boldsymbol{X}(X,g'Y_0) & & \operatorname{Ho}\boldsymbol{X}(X_0,g'Y) \xrightarrow{\psi} \operatorname{Ho}\boldsymbol{Y}(gX_0,Y) \\ h^* \downarrow \quad\quad\quad \downarrow h'_* & \text{and} & h'_* \downarrow \quad\quad\quad \downarrow h^* \\ \operatorname{Ho}\boldsymbol{Y}(fX,Y_0) \xrightarrow{\phi} \operatorname{Ho}\boldsymbol{X}(X,f'Y_0) & & \operatorname{Ho}\boldsymbol{X}(X_0,f'Y) \xrightarrow{\psi} \operatorname{Ho}\boldsymbol{Y}(fX_0,Y) \end{array}$$

commute.

We also note the

43.3. Sufficient conditions for composability of partial adjunctions. Let

$$f_1\colon \boldsymbol{X} \longleftrightarrow \boldsymbol{Y} :f'_1 \quad \text{and} \quad f_2\colon \boldsymbol{Y} \longleftrightarrow \boldsymbol{Z} :f'_2$$

be two composable deformable adjunctions (43.1) such that the pairs (f_1, f_2) and (f'_2, f'_1) are locally left and right deformable (42.3) respectively and let $\boldsymbol{X}_0 \subset \boldsymbol{X}$ and $\boldsymbol{Z}_0 \subset \boldsymbol{Z}$ be left and right deformation retracts on which respectively f_1 and $f_2 f_1$

and f'_2 and $f'_1f'_2$ are homotopical. Then, for every pair of objects $X_0 \in \boldsymbol{X}_0$ and $Z_0 \in \boldsymbol{Z}_0$, the following diagrams commute

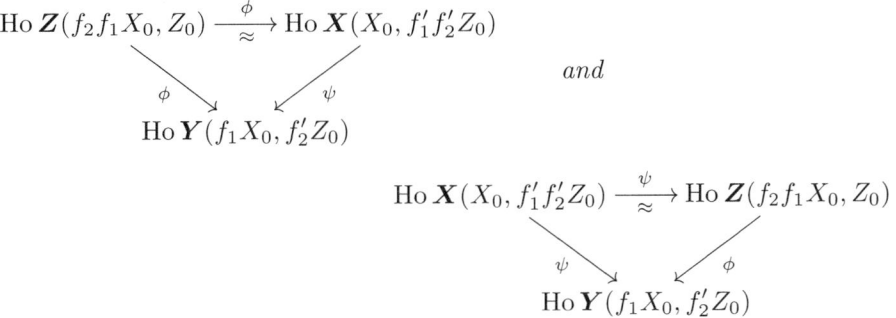

The remainder of this section will be devoted to the outline of a proof of these two propositions, starting with a

43.4. Proof of 43.2. Let η and ε denote the unit and the counit (39.10) of the given adjunction and let (r,s) and (r',s') respectively be a left deformation of $\boldsymbol{X}$ into $\boldsymbol{X}_0$ and a right deformation of $\boldsymbol{Y}$ into $\boldsymbol{Y}_0$ (40.1). Then it follows readily from 33.10 and 39.10 that the functions which, for every pair of objects $X \in \boldsymbol{X}$ and $Y_0 \in \boldsymbol{Y}_0$, and every pair of objects $X_0 \in \boldsymbol{X}_0$ and $Y \in \boldsymbol{Y}$, assign to restricted zigzags (34.2) of the form

$$fX \mathrel{-\!\!\!-\!\!\!-} \cdots \mathrel{-\!\!\!-\!\!\!-} Y_0 \quad \text{in } \boldsymbol{Y} \quad \text{and} \quad X_0 \mathrel{-\!\!\!-\!\!\!-} \cdots \mathrel{-\!\!\!-\!\!\!-} f'Y \quad \text{in } \boldsymbol{X}$$

the restricted zigzags

$$X \xrightarrow{s'\eta} f'r'fX \mathrel{-\!\!\!-\!\!\!-} \cdots \mathrel{-\!\!\!-\!\!\!-} f'r'Y_0 \xleftarrow{s'} f'Y_0 \quad \text{in } \boldsymbol{X}, \quad \text{and}$$

$$fX_0 \xleftarrow{s} frX_0 \mathrel{-\!\!\!-\!\!\!-} \cdots \mathrel{-\!\!\!-\!\!\!-} frf'Y \xrightarrow{\varepsilon s} Y \quad \text{in } \boldsymbol{Y},$$

induce functions

$$\phi\colon \operatorname{Ho}\boldsymbol{Y}(fX, Y_0) \longrightarrow \operatorname{Ho}\boldsymbol{X}(X, f'Y_0) \quad \text{and} \quad \psi\colon \operatorname{Ho}\boldsymbol{X}(X_0, f'Y) \longrightarrow \operatorname{Ho}\boldsymbol{Y}(fX_0, Y)$$

which do not depend on the choice of the deformations (r,s) and (r',s') and which are
  (i) natural in both variables, and
  (ii) compatible with the given adjunction, in the sense of 43.2(i).

To prove 43.2(i) it thus remains to prove the uniqueness of ϕ (and hence by duality of ψ), i.e. we have to show that
  (iii) if ϕ' is a function which, to every pair of objects $X \in \boldsymbol{X}$ and $Y_0 \in \boldsymbol{Y}_0$, assigns a function $\phi'\colon \operatorname{Ho}\boldsymbol{Y}(fX, Y_0) \to \operatorname{Ho}\boldsymbol{X}(X, f'Y_0)$ satisfying (i) and (ii) above, then $\phi' = \phi$.

To prove (iii) we first show that if, for a homotopical category $\boldsymbol{Z}$ and restricted zigzag $Z_0 \mathrel{-\!\!\!-\!\!\!-} \cdots \mathrel{-\!\!\!-\!\!\!-} Z_n$ in $\boldsymbol{Z}$ ($n \geq 0$), we denote by $[\, Z_0 \mathrel{-\!\!\!-\!\!\!-} \cdots \mathrel{-\!\!\!-\!\!\!-} Z_n \,]$ the element of $\operatorname{Ho}\boldsymbol{Z}(Z_0, Z_n)$ "containing" it (33.10), then

(iv) for every restricted zigzag in $\boldsymbol{Y}$ of the form $fX \to Y_1 - \cdots - Y_n$ $(n \geq 1)$ with $Y_1, \cdots, Y_n \in \boldsymbol{Y}_0$, the assumptions of (iii) imply that

$$\phi'[\, fX \to Y_1 - \cdots - Y_n \,] = [\, X \to f'Y_1 - \cdots - f'Y_n \,]$$

where the map $X \to f'Y_1$ is the adjunct of the map $fX \to Y_1$ and the zigzag $f'Y_1 - \cdots - f'Y_n$ is obtained by applying the functor f' to the zigzag $Y_1 - \cdots - Y_n$.

We do this by noting that

(v) in view of (ii) above, (iv) holds for $n = 1$, and

(vi) if, for some integer $k > 0$, (iv) holds for all integers $n \leq k$, then (i) above implies that, for every such zigzag $fX \to Y_1 - \cdots - Y_k$ and map $Y_k \to Y_{k+1} \in \boldsymbol{Y}_0$,

$$\phi'[\, fX \to Y_1 - \cdots - Y_k \to Y_{k+1} \,] = [\, X \to f'Y_1 - \cdots - f'Y_k \to f'Y_{k+1} \,]$$

and that, for every such zigzag $fX \to Y_1 - \cdots - Y_k$ and *weak equivalence* $w \colon Y_{k+1} \to Y_k \in \boldsymbol{Y}_{f'}$, the map

$$[\, X \to f'Y_1 - \cdots - f'Y_k \,] = \phi'[\, fX \to Y_1 - \cdots - Y_k \,]$$
$$= \phi'[\, fX \to Y_1 - \cdots - Y_k \xleftarrow{w} Y_{k+1} \xrightarrow{w} Y_k \,]$$

is the composition of $\phi[\, fX \to Y_1 - \cdots - Y_k \xleftarrow{w} Y_{k+1} \,]$ with the *isomorphism* $\operatorname{Ho} f'w$ which implies that

$$\phi'[\, fX \to Y_1 - \cdots - Y_k \xleftarrow{w} Y_{k+1} \,] = [\, X \to f'Y_1 - \cdots - f'Y_k \xleftarrow{f'w} f'Y_{k+1} \,] \,.$$

The desired result, i.e. (iii), now readily follows from the observation that, for every restricted zigzag of the form $fX - \cdots - Y_0$ with $Y_0 \in \boldsymbol{Y}_0$, the existence of the associated commutative diagram

$$\begin{array}{ccc} fX & - \cdots - & Y_0 \\ {\scriptstyle s'}\downarrow & & \downarrow{\scriptstyle s'} \\ r'fX & - \cdots - & r'Y_0 \end{array}$$

implies that

$$[\, fX - \cdots - Y_0 \,] = [\, fX \xrightarrow{s'} r'fX - \cdots - r'Y_0 \xleftarrow{s'} Y_0 \,] \,.$$

Next we prove 43.2(ii) by noting that, for every pair of objects $X_0 \in \boldsymbol{X}_0$ and $Y_0 \in \boldsymbol{Y}_0$, the function $\psi\phi$ sends, for every restricted zigzag in $\boldsymbol{Y}$ of the form $fX_0 - \cdots - Y_0$, the corresponding element of $\operatorname{Ho} \boldsymbol{Y}(fX_0, Y_0)$ to the element corresponding to the top row, and hence (33.10) the bottom row, of the following commutative diagram (39.10) in which all three rows are restricted zigzags

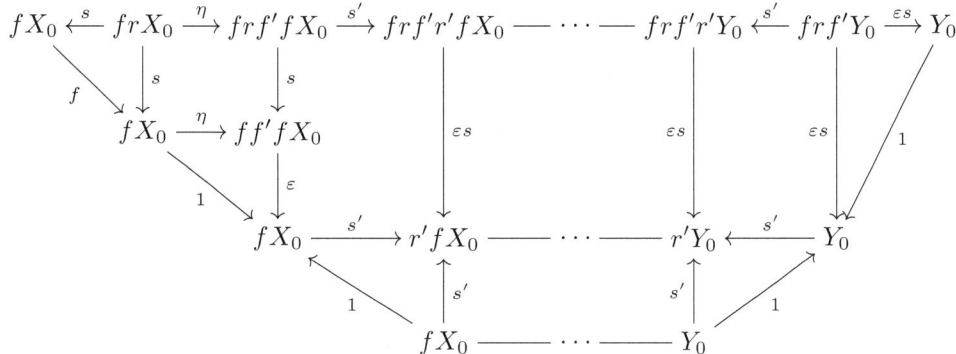

and that a dual assertion holds for the function $\phi\psi$.

It thus remains to prove 43.2(iv). But this now follows from 33.10 and the observation that, in view of the commutative diagram which we mentioned just before 39.10(v), every reduced zigzag in Y of the form $gX - \cdots - Y_0$ gives rise to a commutative diagram in X of the form

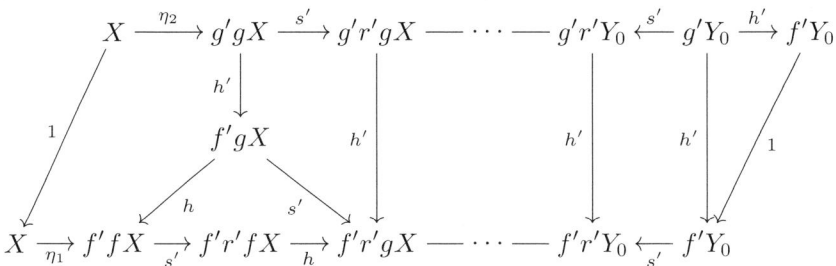

in which η_1 and η_2 denote the units of the two adjunctions.

43.5. Proof of 43.3. The commutativity of the first triangle readily follows from 33.10 and the fact that (39.10) every restricted zigzag in Z of the form

138   VII. DEFORMABLE FUNCTORS AND THEIR APPROXIMATIONS

$f_2 f_1 X_0 - \cdots - Z_0$ gives rise to a commutative diagram in $\boldsymbol{Y}$ of the form

$f_1 X_0 \xleftarrow{s_1} f_1 r_1 X_0 \xrightarrow{\eta_1} f_1 r_1 f_1' f_1 X_0 \xrightarrow{\eta_2} f_1 r_1 f_1' f_2' f_2 f_1 X_0 \xrightarrow{s_2'} f_1 r_1 f_1' f_2' r_2' f_2 f_1 X_0 \longrightarrow \cdots$

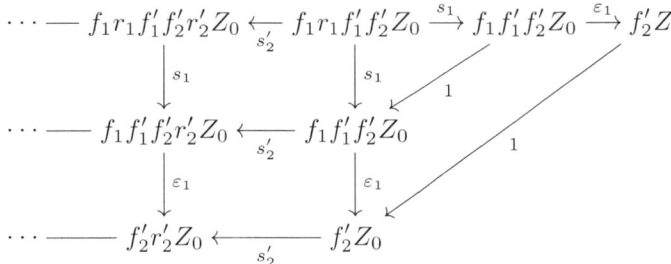

$\cdots \longrightarrow f_1 r_1 f_1' f_2' r_2' Z_0 \xleftarrow{s_2'} f_1 r_1 f_1' f_2' Z_0 \xrightarrow{s_1} f_1 f_1' f_2' Z_0 \xrightarrow{\varepsilon_1} f_2' Z$

$\cdots \longrightarrow f_1 f_1' f_2' r_2' Z_0 \xleftarrow{s_2'} f_1 f_1' f_2' Z_0$

$\cdots \longrightarrow f_2' r_2' Z_0 \xleftarrow{s_2'} f_2' Z_0$

in which η_1 and ε_1 denote the unit and the counit of the first adjunction and η_2 denotes the unit of the second.

The commutativity of the second triangle follows immediately from the commutativity of the first triangle and 43.2(ii).

44. Derived adjunctions

We now use the results of the preceding section on partial adjunctions to show that

    (i) for every deformable adjunction $f\colon \boldsymbol{X} \leftrightarrow \boldsymbol{Y} :f'$ (43.1), its adjunction induces, for every left approximation (k,a) of f and every right approximation (k',a') of f', a *derived adjunction*

$$\operatorname{Ho} k\colon \operatorname{Ho} \boldsymbol{X} \longleftrightarrow \operatorname{Ho} \boldsymbol{Y} :\operatorname{Ho} k'$$

which is *natural* with respect to these approximations,

    (ii) for every two deformable adjunctions

$$f\colon \boldsymbol{X} \longleftrightarrow \boldsymbol{Y} :f' \qquad \text{and} \qquad g\colon \boldsymbol{X} \longleftrightarrow \boldsymbol{Y} :g',$$

every deformable conjugate pair of natural transformations between them (43.1) induces *conjugate* pairs of natural transformations between their derived adjunctions, and

    (iii) for every two composable deformable adjunctions

$$f_1\colon \boldsymbol{X} \longleftrightarrow \boldsymbol{Y} :f_2' \qquad \text{and} \qquad f_2\colon \boldsymbol{Y} \longleftrightarrow \boldsymbol{Z} :f_2'$$

for which the pairs (f_1, f_2) and (f_2', f_1') are respectively locally left and right deformable (42.3), the compositions of their derived adjunctions and the derived adjunctions of their composition are connected by conjugate pairs of natural transformations, which last result will be used to finally complete the proof of 42.6.

In order to formulate these results it is convenient to first introduce

44.1. Canonical natural transformations. *Let $f\colon \boldsymbol{X} \to \boldsymbol{Y}$ be a left (resp. a right) deformable functor (40.2). Then there exists a unique function u (resp. u') which assigns to every pair $\bigl((m,b),(k,a)\bigr)$ of objects in*

$$\bigl(\mathrm{Fun}_{\mathrm{w}}(\boldsymbol{X},\boldsymbol{Y}) \downarrow f\bigr) \qquad \bigl(\text{resp. } \bigl(f \downarrow \mathrm{Fun}_{\mathrm{w}}(\boldsymbol{X},\boldsymbol{Y})\bigr)\bigr)$$

*of which (k,a) is a left (resp. a right) approximation of f (41.1), a **canonical natural transformation***

$$u(m,k)\colon \mathrm{Ho}\, m \longrightarrow \mathrm{Ho}\, k \qquad (\text{resp. } u'(k,m)\colon \mathrm{Ho}\, k \longrightarrow \mathrm{Ho}\, m)$$

such that

(i) *if (m,b) is also a left (resp. a right) approximation of f, then $u(m,k)$ (resp. $u'(k,m)$) is a natural isomorphism,*
(ii) *the function u (resp. u') is natural in both variables, and*
(iii) *for every map*

$$t\colon (m,b) \longrightarrow (k,a) \in \bigl(\mathrm{Fun}_{\mathrm{w}}(\boldsymbol{X},\boldsymbol{Y}) \downarrow f\bigr)$$
$$(\text{resp. } t'\colon (k,a) \longrightarrow (m,b) \in \bigl(f \downarrow \mathrm{Fun}_{\mathrm{w}}(\boldsymbol{X},\boldsymbol{Y})\bigr))$$

one has

$$u(m,k) = \mathrm{Ho}\, t \qquad (\text{resp. } u'(k,m) = \mathrm{Ho}\, t').$$

Proof (of the left half). The composition

$$\bigl(\mathrm{Fun}_{\mathrm{w}}(\boldsymbol{X},\boldsymbol{Y}) \downarrow f\bigr) \xrightarrow{j} \mathrm{Fun}_{\mathrm{w}}(\boldsymbol{X},\boldsymbol{Y}) \xrightarrow{\mathrm{Ho}} \mathrm{Fun}(\mathrm{Ho}\,\boldsymbol{X}, \mathrm{Ho}\,\boldsymbol{Y})$$

in which j denotes the forgetful functor and Ho is as in 33.9, sends weak equivalences to isomorphisms and hence (33.9) admits a unique factorization

$$\bigl(\mathrm{Fun}_{\mathrm{w}}(\boldsymbol{X},\boldsymbol{Y}) \downarrow f\bigr) \xrightarrow{\gamma'} \mathrm{Ho}\bigl(\mathrm{Fun}_{\mathrm{w}}(\boldsymbol{X},\boldsymbol{Y}) \downarrow f\bigr) \longrightarrow \mathrm{Fun}(\mathrm{Ho}\,\boldsymbol{X}, \mathrm{Ho}\,\boldsymbol{Y})$$

in which γ' denotes the localization functor and one now readily verifies that the function which sends a pair $\bigl((m,b),(k,a)\bigr)$ to the image in $\mathrm{Fun}(\mathrm{Ho}\,\boldsymbol{X}, \mathrm{Ho}\,\boldsymbol{Y})$ of the (in view of 38.3(ii)) *unique* map

$$\gamma'(m,b) \longrightarrow \gamma'(k,a) \in \mathrm{Ho}\bigl(\mathrm{Fun}_{\mathrm{w}}(\boldsymbol{X},\boldsymbol{Y}) \downarrow f\bigr)$$

has the desired properties.

Using these canonical natural transformations we now implicitly define

44.2. Derived adjunctions. *Given a deformable adjunction (43.1) $f\colon \boldsymbol{X} \leftrightarrow \boldsymbol{Y} : f'$, there is a unique function which assigns to every pair consisting of a left approximation (k,a) of f (41.1) and a right approximation (k',a') of f', a **derived adjunction***

$$\mathrm{Ho}\, k \colon \mathrm{Ho}\,\boldsymbol{X} \longleftrightarrow \mathrm{Ho}\,\boldsymbol{Y} : \mathrm{Ho}\, k'$$

such that

(i) *for every two left approximations (k_1, a_1) and (k_2, a_2) of f and right approximations (k'_1, a'_1) and (k'_2, a'_2) of f', the canonical natural isomorphisms (44.1)*

$$u(k_1, k_2) \colon \operatorname{Ho} k_1 \longrightarrow \operatorname{Ho} k_2 \quad \text{and} \quad u'(k'_2, k'_1) \colon \operatorname{Ho} k'_2 \longrightarrow \operatorname{Ho} k'_1$$

are conjugate (39.10) with respect to the derived adjunctions

$$\operatorname{Ho} k_1 \colon \operatorname{Ho} \boldsymbol{X} \longleftrightarrow \operatorname{Ho} \boldsymbol{Y} \colon \operatorname{Ho} k'_1 \quad \text{and} \quad \operatorname{Ho} k_2 \colon \operatorname{Ho} \boldsymbol{X} \longleftrightarrow \operatorname{Ho} \boldsymbol{Y} \colon \operatorname{Ho} k'_2$$

and,

(ii) *for every left f-deformation (r, s) (40.2) and right f'-deformation (r', s'), the adjunction isomorphism of the derived adjunction*

$$\operatorname{Ho}(fr) \colon \operatorname{Ho} \boldsymbol{X} \longleftrightarrow \operatorname{Ho} \boldsymbol{Y} \colon \operatorname{Ho}(f'r')$$

associated with the pair $\bigl((fr, fs), (f'r', f's')\bigr)$ (41.2) assigns to every pair of objects $X \in \boldsymbol{X}$ and $Y \in \boldsymbol{Y}$ the composition

$$\operatorname{Ho} \boldsymbol{Y}(frX, Y) \approx \operatorname{Ho} \boldsymbol{Y}(frX, r'Y) \stackrel{\phi}{\approx} \operatorname{Ho} \boldsymbol{X}(rX, f'r'Y) \approx \operatorname{Ho} \boldsymbol{X}(X, f'r'Y)$$

in which ϕ is the partial adjunction isomorphism (43.2) and the other isomorphisms are induced by s' and s.

Proof. In view of 39.10(vi) it suffices to show that, for every two left f-deformations (r_1, s_1) and (r_2, s_2) and right f-deformations (r'_1, s'_1) and (f'_2, s'_2), the canonical natural isomorphisms (44.1)

$$u(fr_1, fr_2) \colon \operatorname{Ho}(fr_1) \longrightarrow \operatorname{Ho}(fr_2) \quad \text{and} \quad u'(f'r'_2, f'r'_1) \colon \operatorname{Ho}(f'r'_2) \longrightarrow \operatorname{Ho}(f'r'_1)$$

are conjugate with respect to the adjunctions

$$\operatorname{Ho}(fr_1) \colon \operatorname{Ho} \boldsymbol{X} \longleftrightarrow \operatorname{Ho} \boldsymbol{Y} \colon \operatorname{Ho}(f'r'_1) \quad \text{and} \quad \operatorname{Ho}(fr_2) \colon \operatorname{Ho} \boldsymbol{X} \longleftrightarrow \operatorname{Ho} \boldsymbol{Y} \colon \operatorname{Ho}(f'r'_2)$$

defined in (ii), but as (40.2, 41.3(iii) and 42.1) the compositions

$$(r_2 r_1, s_2 s_1) \quad \text{and} \quad (r'_2 r'_1, s'_2 s'_1)$$

of respectively (r_1, s_1) and (r_2, s_2) and r'_1, s'_1 and (r'_2, s'_2) (in which $s_2 s_1$ and $s'_2 s'_1$ denote the diagonals of the commutative diagrams

$$\begin{array}{ccc} r_2 r_1 & \xrightarrow{s_2 r_1} & r_1 \\ {\scriptstyle r_2 s_1} \downarrow & & \downarrow {\scriptstyle s_1} \\ r_2 & \xrightarrow{s_2} & 1 \end{array} \quad \text{and} \quad \begin{array}{ccc} 1 & \xrightarrow{s'_2} & r'_2 \\ {\scriptstyle s'_1} \downarrow & & \downarrow {\scriptstyle r'_2 s'_1} \\ r'_1 & \xrightarrow{s'_2 r'_1} & r'_2 r'_1 \end{array} \quad)$$

are a left f-deformation and a right f'-deformation, this follows readily from the commutativity of the diagram

$$\begin{array}{ccccccc}
\operatorname{Ho}\boldsymbol{Y}(fr_2X,Y) & \approx & \cdots & \stackrel{\phi}{\approx} & \cdots & \approx & \operatorname{Ho}\boldsymbol{X}(X,f'r_2'Y) \\
{\scriptstyle (fr_2s_1)^*}\downarrow & & & & & & \downarrow {\scriptstyle (f'r_2's_1')_*} \\
\operatorname{Ho}\boldsymbol{Y}(fr_2r_1X,Y) & \approx & \cdots & \stackrel{\phi}{\approx} & \cdots & \approx & \operatorname{Ho}\boldsymbol{X}(X,f'r_2'r_1'Y) \\
{\scriptstyle (fs_2r_1)^*}\uparrow & & & & & & \uparrow {\scriptstyle (f's_2'r_1')_*} \\
\operatorname{Ho}\boldsymbol{Y}(fr_1X,Y) & \approx & \cdots & \stackrel{\phi}{\approx} & \cdots & \approx & \operatorname{Ho}\boldsymbol{X}(X,f'r_1'Y)
\end{array}$$

in which the horizontal sequences are as in (ii), and the observation that (44.1)

$$\operatorname{Ho}(fr_2s_1)\operatorname{Ho}(fs_2r_1)^{-1} = u(fr_1,fr_2)$$

and

$$\operatorname{Ho}(f's_2'r_1')^{-1}\operatorname{Ho}(f'r_2's_1') = u'(f'r_2',f'r_1') \ .$$

Next we consider

44.3. Conjugations between derived adjunctions. *Let*

$$f\colon \boldsymbol{X} \longleftrightarrow \boldsymbol{Y} :f' \qquad \text{and} \qquad g\colon \boldsymbol{X} \longleftrightarrow \boldsymbol{Y} :g'$$

be deformable adjunctions, let

$$h\colon f \longrightarrow g \qquad \text{and} \qquad h'\colon g' \longrightarrow f'$$

be a deformable conjugate pair of natural transformations between them (43.1), and let

$$k\colon (k_1,a_1) \longrightarrow (k_2,a_2) \qquad \text{and} \qquad k'\colon (k_2',a_2') \longrightarrow (k_2',a_1')$$

respectively be a left approximation of h and a right approximation of h' (41.6). Then the natural transformations

$$\operatorname{Ho}k\colon \operatorname{Ho}k_1 \longrightarrow \operatorname{Ho}k_2 \qquad \text{and} \qquad \operatorname{Ho}k'\colon \operatorname{Ho}k_2' \longrightarrow \operatorname{Ho}k_1'$$

are conjugate with respect to the derived adjunction (44.2)

$$\operatorname{Ho}k_1\colon \operatorname{Ho}\boldsymbol{X} \longleftrightarrow \operatorname{Ho}\boldsymbol{Y} :\operatorname{Ho}k_1' \qquad \text{and} \qquad \operatorname{Ho}k_2\colon \operatorname{Ho}\boldsymbol{X} \longleftrightarrow \operatorname{Ho}\boldsymbol{Y} :\operatorname{Ho}k_2' \ .$$

Proof. This follows from 39.10(vi), 44.2(i) and the observation that, in view of 43.2(iii), for every left h-deformation (r,s) and right h'-deformation (r',s') (40.6) and every pair of objects $X \in \boldsymbol{X}$ and $Y \in \boldsymbol{Y}$, the diagram

$$\begin{array}{ccccccc}
\operatorname{Ho}\boldsymbol{Y}(grX,Y) & \approx & \operatorname{Ho}\boldsymbol{Y}(grX,r'Y) & \stackrel{\phi}{\approx} & \operatorname{Ho}\boldsymbol{X}(rX,g'r'Y) & \approx & \operatorname{Ho}\boldsymbol{X}(X,g'r'Y) \\
{\scriptstyle h^*}\downarrow & & {\scriptstyle h^*}\downarrow & & \downarrow{\scriptstyle h'_*} & & \downarrow{\scriptstyle h'_*} \\
\operatorname{Ho}\boldsymbol{Y}(frX,Y) & \approx & \operatorname{Ho}\boldsymbol{Y}(frX,r'Y) & \stackrel{\phi}{\approx} & \operatorname{Ho}\boldsymbol{X}(rX,f'r'Y) & \approx & \operatorname{Ho}\boldsymbol{X}(X,f'r'Y)
\end{array}$$

in which the horizontal sequences are as in 44.2(ii), commutes.

Finally we discuss

44.4. Compositions of derived adjunctions. Let
$$f_1\colon X \longleftrightarrow Y :f_1' \qquad\text{and}\qquad f_2\colon Y \longleftrightarrow Z :f_2'$$
be deformable adjunctions (43.1) for which the pairs (f_1, f_2) and (f_2', f_1') are respectively locally left and right deformable (42.3). Then

(i) *their composition*
$$f_2 f_1\colon X \longleftrightarrow Z :f_1' f_2'$$
is also a deformable adjunction.

Moreover, if
$$(k_1, a_1), \quad (k_2, a_2) \quad\text{and}\quad (k_{21}, a_{21})$$
are left approximations of f_1, f_2 and $f_2 f_1$ and
$$(k_1', a_1'), \quad (k_2', a_2') \quad\text{and}\quad (k_{12}', a_{12}')$$
are right approximations of f_1', f_2' and $f_1' f_2'$ respectively (41.1), then

(ii) *the canonical natural transformations* (44.1)
$$u(k_2 k_1, k_{21})\colon \mathrm{Ho}(k_2 k_1) \longrightarrow \mathrm{Ho}\, k_{21} \qquad\text{and}\qquad u'(k_{12}', k_1' k_2')\colon \mathrm{Ho}\, k_{12}' \longrightarrow \mathrm{Ho}(k_1' k_2')$$
associated with the composition (42.1) $(k_2 k_1, a_2 a_1)$ of (k_1, a_1) and (k_2, a_2) and the left approximation (k_{21}, a_{21}) and with the composition $(k_1' k_2', a_1' a_2')$ of (k_2', a_2') and (k_1', a_1') and the right approximation (k_{12}', a_{12}'), are conjugate with respect to the derived adjunction (44.2)
$$\mathrm{Ho}(k_{21})\colon \mathrm{Ho}\, X \longleftrightarrow \mathrm{Ho}\, Z :\mathrm{Ho}(k_{12}')$$
and the composition of the derived adjunctions
$$\mathrm{Ho}\, k_1\colon \mathrm{Ho}\, X \longleftrightarrow \mathrm{Ho}\, Y :\mathrm{Ho}\, k_1' \qquad\text{and}\qquad \mathrm{Ho}\, k_2\colon \mathrm{Ho}\, Y \longleftrightarrow \mathrm{Ho}\, Z :\mathrm{Ho}\, k_2'\ .$$

44.5. Corollary. Let
$$f_1\colon X \longleftrightarrow Y :f_1' \qquad\text{and}\qquad f_2\colon Y \longleftrightarrow Z :f_2'$$
be deformable adjunctions for which the pairs (f_1, f_2) and (f_2', f_1') are respectively left and right deformable (42.3). Then every composition of a derived adjunction of the first adjunction with a derived adjunction of the second adjunction is a derived adjunction of their composition.

Proof. In view of 39.10(vi), 44.1 and 44.2 it suffices to prove that if
$$(r_1, s_1), \quad (r_2, s_2), \quad (r_1', s_1') \quad\text{and}\quad (r_2', s_2')$$
respectively are a left f_1- as well as a left $f_2 f_1$-deformation, a left f_2-deformation, a right f_1'-deformation and a right f_2'- as well as a right $f_1' f_2'$-deformation, then the natural transformations
$$\mathrm{Ho}(f_2 s_2 f_1 r_1)\colon \mathrm{Ho}(f_2 r_2 f_1 r_1) \longrightarrow \mathrm{Ho}(f_2 f_1 r_1) \qquad\text{and}$$
$$\mathrm{Ho}(f_1' s_1' f_2' r_2')\colon \mathrm{Ho}(f_1' f_2' r_2') \longrightarrow \mathrm{Ho}(f_1' r_1' f_2' r_2')$$
are conjugate with respect to the derived adjunction
$$\mathrm{Ho}(f_2 f_1 r)\colon \mathrm{Ho}\, X \longleftrightarrow \mathrm{Ho}\, Z :\mathrm{Ho}(f_1' f_2' r_2')$$

and the composition of the derived adjunctions

$\operatorname{Ho}(f_1 r_1)\colon \operatorname{Ho} \boldsymbol{X} \longleftrightarrow \operatorname{Ho} \boldsymbol{Y} : \operatorname{Ho}(f_1' r_1')$ and $\operatorname{Ho}(f_2 r_2)\colon \operatorname{Ho} \boldsymbol{Y} \longleftrightarrow \operatorname{Ho} \boldsymbol{Z} : \operatorname{Ho}(f_2' r_2')$

But this follows readily from the observation that, in view of 43.3 and the naturality of the partial adjunction functions ϕ and ψ (43.2(i)) and of the maps induced by s_1, s_2, s_1' and s_2', for every pair of objects $X \in \boldsymbol{X}$ and $Z \in \boldsymbol{Z}$, the following diagram commutes

$$\begin{array}{ccc}
\operatorname{Ho} \boldsymbol{Z}(f_2 r_2 f_1 r_1 X, Z) \xrightarrow[\approx]{(s_2')_*} \operatorname{Ho} \boldsymbol{Z}(f_2 r_2 f_1 r_1 X, r_2' Z) & \xrightarrow{\phi}_{\approx} & \operatorname{Ho} \boldsymbol{Y}(r_2 f_1 r_1 X, f_2' r_2' Z) \\
s_2^* \uparrow \qquad\qquad s_2^* \uparrow & & \uparrow s_2^* \approx \\
\operatorname{Ho} \boldsymbol{Z}(f_2 f_1 r_1 X, Z) \xrightarrow[\approx]{(s_2')_*} \operatorname{Ho} \boldsymbol{Z}(f_2 f_1 r_1 X, r_2' Z) & \xrightarrow{\phi} & \operatorname{Ho} \boldsymbol{Y}(f_1 r_1 X, f_2' r_2' Z) \\
\phi \downarrow \approx & & \downarrow (s_1')_* \approx \\
\operatorname{Ho} \boldsymbol{X}(X, f_1' f_2' r_2' Z) \xrightarrow[s_1^*]{\approx} \operatorname{Ho} \boldsymbol{X}(r_1 X, f_1' f_2' r_2' Z) & \xrightarrow{\psi} & \operatorname{Ho} \boldsymbol{Y}(f_1 r_1 X, r_1' f_2' r_2' Z) \\
(s_1')_* \downarrow \qquad\qquad (s_1')_* \downarrow & & \\
\operatorname{Ho} \boldsymbol{X}(X, f_1' r_1' f_2' r_2' Z) \xrightarrow[s_1^*]{\approx} \operatorname{Ho} \boldsymbol{X}(r_1 X, f_1' r_1' f_2' r_2' Z) & \xrightarrow{\psi} &
\end{array}$$

45. The Quillen condition

As another application of the results of 44.2 we show in this last section of the chapter

  (i) that a *sufficient* condition on a deformable adjunction (43.1) in order that the left approximations of the left adjoints (41.1) and the right approximations of the right adjoint are homotopically inverse homotopical equivalences of homotopical categories (33.1) is that the adjoint pair satisfies a so-called *Quillen condition*, and
  (ii) that, if the homotopical categories in question are saturated (33.9), then this condition is also *necessary*.

In more detail

45.1. The Quillen condition. A deformable adjunction (43.1) $f\colon \boldsymbol{X} \leftrightarrow \boldsymbol{Y} : f'$ will be said to satisfy the **Quillen condition** if

  (i) for *some* left f-deformation retract $\boldsymbol{X}_0 \subset \boldsymbol{X}$ (40.2) and right f'-deformation retract $\boldsymbol{Y}_0 \subset \boldsymbol{Y}$ and *every* pair of objects $X_0 \in \boldsymbol{X}_0$ and $Y_0 \in \boldsymbol{Y}_0$, a map $fX_0 \to Y_0 \in \boldsymbol{Y}$ is a weak equivalence *iff* its adjoint $X_0 \to f'Y_0 \in \boldsymbol{X}$ is so.

which one readily verifies (using 40.3) is equivalent to the seemingly stronger requirement that

  (ii) for *every* left f-deformation retract $\boldsymbol{X}_0 \subset \boldsymbol{X}$ and right f'-deformation retract $\boldsymbol{Y}_0 \subset \boldsymbol{Y}$ and *every* pair of objects $X_0 \in \boldsymbol{X}_0$ and $Y_0 \in \boldsymbol{Y}_0$, a map $fX_0 \to Y_0 \in \boldsymbol{Y}$ is a weak equivalence *iff* its adjunct $X_0 \to f'Y_0 \in \boldsymbol{X}$ is so.

The usefulness of this condition is due to the following

45.2. Consequences of the Quillen condition. *Given a deformable adjunctions (43.1)* $f\colon \mathbf{X} \leftrightarrow \mathbf{Y} :f'$, *a left approximation* (k,a) *of* f *and a right approximation* (k',a') *of* f' *(41.1), consider the following statements:*

  (i) *the adjunction satisfies the Quillen condition (45.1),*
  (ii) *the functors k and k' are homotopically inverse homotopical equivalences of homotopical categories (33.1),*
  (ii)′ *the functor k is a homotopical equivalence of homotopical categories (33.1),*
  (ii)″ *the functor k' is a homotopical equivalence of homotopical categories,*
  (iii) *the functors $\operatorname{Ho} k$ and $\operatorname{Ho} k'$ are inverse equivalences of categories,*
  (iii)′ *the functor $\operatorname{Ho} k$ is a equivalence of categories, and*
  (iii)″ *the functor $\operatorname{Ho} k'$ is an equivalence of categories.*

Then

  (iv) *the following implications hold*

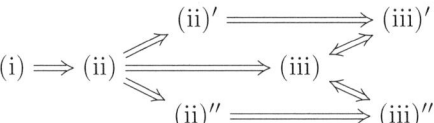

Moreover

  (v) *if $\mathbf{X}$ and $\mathbf{Y}$ are saturated (33.8), then each of (iii), (iii)′ and (iii)″ implies (i), and hence each of the seven statements (i)–(iii)″ implies all the others.*

Proof. In view of the homotopical uniqueness of approximations (41.1), this result follows readily from 44.2 and

45.3. Proposition. *Given a deformable adjunction (43.1)* $f\colon \mathbf{X} \leftrightarrow \mathbf{Y} :f'$, *a left f-deformation (r,s) (40.2) and a right f'-deformation (r',s'), consider the following three statements:*

  (i) *for every pair of objects $X_0 \in \mathbf{X}$ and $Y_0 \in \mathbf{Y}$ which are in the images of r and r' respectively, a map $X_0 \to f'Y_0 \in \mathbf{X}$ is a weak equivalence if (resp. only if) its adjunct $fX_0 \to Y_0 \in \mathbf{Y}$ is so,*
  (ii) *the zigzag of natural transformations*

$$1_{\mathbf{X}} \xleftarrow{s} r \longrightarrow f'r'fr \qquad (\text{resp. } frf'r' \longrightarrow r' \xleftarrow{s'} 1_{\mathbf{Y}})$$

  in which the unnamed map is the adjunct of the natural weak equivalence

$$fr \xrightarrow{s'fr} r'fr \qquad (\text{resp. } rf'r' \xrightarrow{sf'r'} f'r')$$

  is a zigzag of natural weak equivalences, and
  (iii) *the unit (resp. the counit) (39.10)*

$$1_{\operatorname{Ho}\mathbf{X}} \longrightarrow \operatorname{Ho}(f'r')\operatorname{Ho}(fr) \qquad (\text{resp. } \operatorname{Ho}(fr)\operatorname{Ho}(f'r') \longrightarrow 1_{\operatorname{Ho}\mathbf{Y}})$$

  of the derived adjunction $\operatorname{Ho}(fr)\colon \operatorname{Ho}\mathbf{X} \leftrightarrow \operatorname{Ho}\mathbf{Y} :\operatorname{Ho}(f'r')$ (44.2) is a natural isomorphism.

Then (i) implies (ii) and (ii) implies (iii). Moreover, if the homotopical category $\mathbf{X}$ (resp. $\mathbf{Y}$) is saturated (33.9), then (iii) implies (i) and hence (i), (ii) and (iii) are equivalent statements.

Proof (of the first half). Clearly (i) implies (ii) and a straightforward calculation using 43.2(i) and 44.2 yields that the image in Ho $\boldsymbol{X}$ of the zigzag $1_{\boldsymbol{X}} \xleftarrow{s} r \to f'r'fr$ is exactly the unit $1_{\mathrm{Ho}\,\boldsymbol{X}} \to \mathrm{Ho}(f'r)\,\mathrm{Ho}(fr)$, so that (ii) implies (iii).

It thus remains to show that, if $\boldsymbol{X}$ is saturated, then (iii) implies (i) and we do this by successively noting that if $p\colon X_0 \to f'Y_0 \in \boldsymbol{X}$ is a map such that its adjunct $q\colon fX_0 \to Y_0 \in \boldsymbol{Y}$ is a weak equivalence, then

  (i) the map $q(fsX_0)\colon frX_0 \to Y_0 \in \boldsymbol{Y}$ is also a weak equivalence and hence its image $\gamma'\bigl(q(fsX_0)\bigr) \in \mathrm{Ho}\,\boldsymbol{Y}$ under the localization functor $\gamma'\colon \boldsymbol{Y} \to \mathrm{Ho}\,\boldsymbol{Y}$ is an isomorphism,

  (ii) in view of 43.2(i) and 44.2 the adjunct of $\gamma'\bigl(q(fsX_0)\bigr)$ is the image $\gamma\bigl((f's'Y_0)p\bigr) \in \mathrm{Ho}\,\boldsymbol{X}$ of the map $(f's'Y_0)p\colon X_0 \to f'r'Y_0 \in \boldsymbol{X}$ under the localization functor $\gamma\colon \boldsymbol{X} \to \mathrm{Ho}\,\boldsymbol{X}$,

  (iii) this adjunct $\gamma\bigl((f's'Y_0)p\bigr)$ is the composition of the unit (which is assumed to be an isomorphism) with the image of $\gamma'\bigl(q(fsX_0)\bigr)$ under the functor $\mathrm{Ho}(f'r')$ and hence is also an isomorphism, and

  (iv) in view of the saturation of $\boldsymbol{X}$, the map $(f's'Y_0)p$ is thus a weak equivalence and so is therefore, because of the two out of three property, the original map $p\colon X_0 \to fY_0 \in \boldsymbol{X}$.

CHAPTER VIII

Homotopy Colimit and Limit Functors and Homotopical Ones

46. Introduction

46.1. Summary. This last chapter consists of two parts.

In the first part (i.e. §§47–49), motivated by the homotopy colimit and limit functors on model categories (see chapter IV), we define *homotopy colimit* and *limit functors* on arbitrary cocomplete and complete homotopical categories and obtain sufficient conditions for their *existence* and *composability* as well as sufficient conditions for the *homotopical cocompleteness* and *completeness* of such homotopical categories, in a sense which is much stronger than one would expect.

In the second part (i.e. §§50 and 51) we define, on *not necessarily* cocomplete and complete homotopical categories, *homotopical colimit* and *limit functors* which, in the cocomplete and complete case, turn out to be essentially the same as the above homotopy colimit and limit functors, and note that there are corresponding notions of homotopical cocompleteness and completeness which, in the cocomplete and complete case, reduce to the ones mentioned above.

46.2. Some of the tools. To deal with homotopy colimit and limit functors we need the results of chapter VII on the *approximations* of (not necessarily homotopical) functors between homotopical categories and of composable pairs of such functors, but in order to be able to deal with the associated notions of homotopical cocompleteness and completeness we first have to generalize these results on composable pairs of such functors to diagram-like collections of homotopical categories and such functors between them, indexed by a category $\boldsymbol{C}$, which we will call *left* and *right* $\boldsymbol{C}$-*systems* and which are functions F which assign

(i) to every object $D \in \boldsymbol{C}$ a homotopical category $F\boldsymbol{D}$,
(ii) to every map $u\colon A \to B \in \boldsymbol{C}$ a (not necessarily homotopical) functor $Fu\colon FA \to FB$, and
(iii) to every composable pair of maps $u\colon A \to B$ and $v\colon B \to D \in \boldsymbol{C}$, a *natural weak equivalence*

$$F(v,u)\colon (Fv)(Fu) \longrightarrow F(vu) \quad \text{or} \quad F(v,u)\colon F(vu) \longrightarrow (Fv)(Fu)$$

which is associative in the obvious sense.

The treatment of homotopical colimit and limit functors and the associated notions of homotopical cocompleteness and completeness requires in addition a discussion of *Kan extensions* and *homotopical Kan extensions* of the identity functor along a (not necessarily homotopical) functor between homotopical categories and of the more general notions of Kan extensions and homotopical Kan extensions along a *system*.

46.3. Organization of the chapter. After a warning on the use of the term *adjoint* (in 46.4), we discuss *homotopy colimit* and *limit functors* (in §47), introduce *left* and *right systems* (in §48) and use these to describe the associated notions of *homotopical cocompleteness* and *completeness* (in §49). Next we investigate *Kan extensions* and *homotopical Kan extensions* of identity functors and use these to deal with *homotopical colimit* and *limit functors* (in §50). And in the last section (§51) we introduce the *Kan extensions* and *homotopical Kan extensions* along *systems* which we need to deal with the more general notions of *homotopical cocompleteness* and *completeness*.

We thus end with a

46.4. Warning on the term adjoint. Given a functor $f\colon X \to Y$, we will call a functor $g\colon Y \to X$ a left (or a right) adjoint of f if there is *given* an adjunction

$$g\colon Y \longleftrightarrow X : f \qquad (\text{or } f\colon X \longleftrightarrow Y : g)$$

and *not*, as is done by some authors, if one only assumes the *existence* of such an adjunction.

47. Homotopy colimit and limit functors

In this section, assuming the existence of colimit and limit functors, we

(i) define *homotopy colimit* and *limit functors* on a homotopical category respectively as left and right approximations of an *arbitrary but fixed* such colimit or limit functor, and

(ii) describe sufficient conditions for their *existence, composability* and *homotopical compatibility with deformable adjoints*.

We thus start with defining

47.1. Homotopy colimit and limit functors. Given a homotopical category X and a category D, we

(i) define a **D-colimit** (resp. **D-limit**) **functor** on X as a left (resp. a right) adjoint (46.4) $X^D \to X$ of the **constant diagram functor** $c^*\colon X \to X^D$ (which sends every object of X to the corresponding constant D-diagram), and

(ii) if such a D-colimit (resp. D-limit) functor on X exists, denote by

$$\operatorname{colim}^D \qquad (\text{resp. } \lim^D)$$

an *arbitrary but fixed* such D-colimit (resp. D-limit) functor, and define a **homotopy D-colimit** (resp. **D-limit**) **functor** on X as a left (resp. a right) approximation (41.1) of colim^D (resp. $\lim^D$),

and more generally, given a functor $u\colon A \to B$, we

(i)' define a **u-colimit** (resp. **u-limit**) **functor** on X as a left (resp. a right) adjoint $X^A \to X^B$ of the **induced diagram functor**

$$X^u = u^*\colon X^B \longrightarrow X^A$$

(which sends every functor $B \to X$ to its composition with u), and

(ii)′ if such a *u*-colimit (resp. *u*-limit) functor on $\mathbf{X}$ exists, denote by
$$\operatorname{colim}^u \qquad (\text{resp. } \lim{}^u)$$
an *arbitrary but fixed* such *u*-colimit (resp. *u*-limit) functor, and define a **homotopy *u*-colimit** (resp. ***u*-limit**) **functor** on $\mathbf{X}$ as a left (resp. a right) approximation of colim^u (resp. $\lim^u$).

Then one has the following

47.2. Sufficient conditions for existence of homotopy colimit and limit functors and their derived adjunctions. *For every homotopical category $\mathbf{X}$ and functor $u\colon \mathbf{A} \to \mathbf{B}$*

  (i) *such homotopy u-colimit (resp. u-limit) functors on $\mathbf{X}$, if they exist, are homotopically unique (37.5), and*
  (ii) *a sufficient condition for their existence is that the functor colim^u (resp. $\lim^u$) exist (47.1) and be left (resp. right) deformable (40.2).*

Moreover in that case

  (iii) *every homotopy u-colimit (resp. u-limit) functor on $\mathbf{X}$, i.e. left (resp. right) approximation (k,a) of colim^u (resp. $\lim^u$), comes with the derived adjunction (44.2)*

$$\operatorname{Ho} k\colon \operatorname{Ho} \mathbf{X}^{\mathbf{A}} \longleftrightarrow \operatorname{Ho} \mathbf{X}^{\mathbf{B}} \colon \operatorname{Ho} u^* \qquad (\text{resp. } \operatorname{Ho} u^*\colon \operatorname{Ho} \mathbf{X}^{\mathbf{B}} \longleftrightarrow \operatorname{Ho} \mathbf{X}^{\mathbf{A}} \colon \operatorname{Ho} k)$$

associated with the approximations
$$\big((k,a),(u^*,1_{u_*})\big) \qquad (\text{resp. } \big((u^*,1_{u^*}),(k,a)\big))$$
which has as its counit (resp. unit) the natural transformation

$$\operatorname{Ho}\big(e(au^*)\big)\colon (\operatorname{Ho} k)(\operatorname{Ho} u^*) \longrightarrow 1_{\operatorname{Ho} \mathbf{X}^{\mathbf{B}}}$$
$$(\text{resp. } \operatorname{Ho}\big((au^*)e\big)\colon 1_{\operatorname{Ho} \mathbf{X}^{\mathbf{B}}} \longrightarrow (\operatorname{Ho} k)(\operatorname{Ho} u^*))$$

where (46.4)
$$e\colon \operatorname{colim}^u u^* \longrightarrow 1_{\mathbf{X}^{\mathbf{B}}} \qquad (\text{resp. } e\colon 1_{\mathbf{X}^{\mathbf{B}}} \longrightarrow \lim{}^u u^*)$$
denotes the counit (resp. the unit) of the adjunction
$$\operatorname{colim}^u\colon \mathbf{X}^{\mathbf{A}} \longleftrightarrow \mathbf{X}^{\mathbf{B}} \colon u^* \qquad (\text{resp. } u^*\colon \mathbf{X}^{\mathbf{B}} \longleftrightarrow \mathbf{X}^{\mathbf{A}} \colon \lim{}^u) \ .$$

Proof. Parts (i) and (ii) follow from 41.1(iii) and 41.2(ii), while the colimit half of part (iii) follows from 44.2 and the observation that if (r,s) is a left colim^u-deformation (40.2), then the resulting commutative diagram

$$\begin{array}{ccccc}
(kr)u^* & \xrightarrow{(ks)u^*} & ku^* & & \\
{\scriptstyle (ar)u^*}\downarrow & & {\scriptstyle au^*}\downarrow & \searrow^{e(au^*)} & \\
(\operatorname{colim}^u r)u^* & \xrightarrow{(\operatorname{colim}^u s)u^*} & \operatorname{colim}^u u^* & \xrightarrow{e} & 1_{\mathbf{X}^{\mathbf{B}}}
\end{array}$$

gives rise to the commutative diagram

$$\begin{array}{ccc}
(\operatorname{Ho} kr)(\operatorname{Ho} u^*) & \longrightarrow & (\operatorname{Ho} k)(\operatorname{Ho} u^*) \\
\downarrow & & \downarrow{\scriptstyle \operatorname{Ho}(e(au^*))} \\
(\operatorname{Ho}\operatorname{colim}^u r)(\operatorname{Ho} u^*) & \longrightarrow & 1_{\operatorname{Ho} \mathbf{X}^{\mathbf{B}}}
\end{array}$$

and the first part of the proposition now follows from the observation that, in view of 44.1 and 44.2, the natural transformation $(\operatorname{Ho} k)(\operatorname{Ho} u^*) \to 1_{\operatorname{Ho} \boldsymbol{X}^{\boldsymbol{B}}}$ obtained by going counter clockwise around this diagram is exactly the desired counit.

47.3. Compositions. Given a homotopical category $\boldsymbol{X}$ and two composable functors $u\colon \boldsymbol{A} \to \boldsymbol{B}$ and $v\colon \boldsymbol{B} \to \boldsymbol{D}$ for which there exist u- and v-colimit (resp. u- and v-limit) functors on $\boldsymbol{X}$, the *composition* of a homotopy u-colimit (resp. u-limit) functor on $\boldsymbol{X}$ with a homotopy v-colimit (resp. v-limit) functor will be a slight *modification* (by a natural isomorphism) of the *composition of approximations* which we considered in §42. More precisely, for every homotopy u-colimit (resp. u-limit) functor (k_u, a_u) and homotopy v-colimit (resp. v-limit) functor (k_v, a_v) on $\boldsymbol{X}$, their "**composition**" will be the pair
$$(k_v k_u, \operatorname{colim}^{(v,u)} a_v a_u) \in \left(\operatorname{Fun}_{\mathrm{w}}(\boldsymbol{X}^{\boldsymbol{A}}, \boldsymbol{X}^{\boldsymbol{D}}) \downarrow \operatorname{colim}^{vu}\right)$$
$$(\text{resp.} \quad (k_v k_u, \lim^{(v,u)} a_v a_u) \in \left(\lim^{vu} \downarrow \operatorname{Fun}_{\mathrm{w}}(\boldsymbol{X}^{\boldsymbol{A}}, \boldsymbol{X}^{\boldsymbol{D}})\right) \quad)$$
where $a_v a_u$ denotes the diagonal of the commutative diagram

$$\begin{array}{ccc} k_v k_u & \xrightarrow{a_v} & \operatorname{colim}^v k_u \\ a_u \downarrow & & \downarrow a_u \\ k_u \operatorname{colim}^v & \xrightarrow{a_v} & \operatorname{colim}^v \operatorname{colim}^u \end{array} \quad (\text{resp.} \quad \begin{array}{ccc} \lim^v \lim^u & \xrightarrow{a_v} & k_v \lim^u \\ a_u \downarrow & & \downarrow a_u \\ \lim^v k_u & \xrightarrow{a_v} & k_v k_u \end{array} \quad)$$

and
$$\operatorname{colim}^{(v,u)}\colon \operatorname{colim}^v \operatorname{colim}^u \longrightarrow \operatorname{colim}^{vu} \quad (\text{resp.} \ \lim^{(v,u)}\colon \lim^{vu} \longrightarrow \lim^v \lim^u)$$
is the natural isomorphism which is the conjugate of the identity natural transformation of the functor
$$(vu)^* = u^* v^* \colon \boldsymbol{X}^{\boldsymbol{D}} \longrightarrow \boldsymbol{X}^{\boldsymbol{A}} \ .$$

This pair need not again be a homotopy colimit (resp. limit) functor. However one has the following

47.4. Sufficient conditions for composability of homotopy colimit and limit functors and their derived adjunctions. *A sufficient condition in order that, given a homotopical category $\boldsymbol{X}$ and two composable functors $u\colon \boldsymbol{A} \to \boldsymbol{B}$ and $v\colon \boldsymbol{B} \to \boldsymbol{D}$ for which there exist u- and v-colimit (resp. u- and v-limit) functors on $\boldsymbol{X}$, for every homotopy u-colimit (resp. u-limit) functor (k_u, a_u) and homotopy v-colimit (resp. v-limit) functor (k_v, a_v) on $\boldsymbol{X}$, their (47.3) "composition" is a homotopy vu-colimit (resp. vu-limit) functor on $\boldsymbol{X}$ is that*

(i) *the pair*
$$(\operatorname{colim}^u, \operatorname{colim}^v) \qquad (\text{resp. } (\lim^u, \lim^v))$$
is left (resp. right) deformable (42.3)

which in particular is the case if

(ii) $\boldsymbol{X}$ *is saturated (33.9) and the pair*
$$(\operatorname{colim}^u, \operatorname{colim}^v) \qquad (\text{resp. } (\lim^u, \lim^v))$$
is locally left (resp. right) deformable (42.3).

Moreover, if (i) *holds, then*

(iii) *the composition of the derived adjunctions (47.2) of (k_u, a_u) and (k_v, a_v) is exactly the derived adjunction of their "composition" (47.3).*

Proof. Part (i) follows readily from 42.4 and the fact that $\operatorname{colim}^{(v,u)}$ (resp. $\lim^{v,u}$) is a natural isomorphism, while (ii) follows from 33.9(v) and 42.5 and the observation that, as u^* and v^* are both homotopical functors, the pair (v^*, u^*) is right (resp. left) deformable (42.3).

To prove the colimit half of (iii) one notes that in the commutative diagram

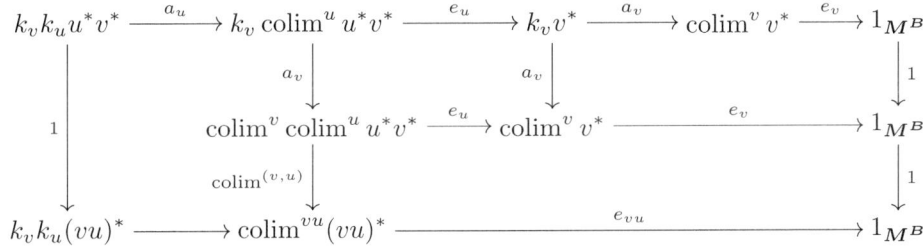

in which e_u, e_v and e_{vu} are (induced by) the relevant counits, the top row, in view of 39.10(iii) and (iv) and 47.2, is the counit of the composition of the derived adjunction of (k_u, a_u) and (k_v, a_v), while (47.2) the bottom row is the counit of the derived adjunction of their "composition".

It thus remains to discuss

47.5. Homotopical compatibility of homotopy colimit and limit functors with deformable adjoints. This will be a homotopical version of the following categorical notion.

Given an adjunction $f\colon \boldsymbol{X} \leftrightarrow \boldsymbol{Y} :g$ and a functor $u\colon \boldsymbol{A} \to \boldsymbol{B}$, we will say that f (resp. g) is **compatible** with u-colimit (resp. u-limit) functors if

(i) there exist u-colimit (resp. u-limit) functors (47.1) on both $\boldsymbol{X}$ and $\boldsymbol{Y}$

as in that case

(ii) *for every pair of u-colimit (resp. u-limit) functors s on $\boldsymbol{X}$ and t on $\boldsymbol{Y}$, the compositions*

$$f^{\boldsymbol{B}}s \quad \text{and} \quad tf^{\boldsymbol{A}}\colon \boldsymbol{X}^{\boldsymbol{A}} \longrightarrow \boldsymbol{Y}^{\boldsymbol{B}} \qquad (\text{resp.} \quad g^{\boldsymbol{B}}t \quad \text{and} \quad sg^{\boldsymbol{A}}\colon \boldsymbol{Y}^{\boldsymbol{A}} \longrightarrow \boldsymbol{X}^{\boldsymbol{B}})$$

are left (resp. right) adjoints of the composition

$$\boldsymbol{X}^u g^{\boldsymbol{B}} = g^{\boldsymbol{A}} \boldsymbol{Y}^u \colon \boldsymbol{Y}^{\boldsymbol{B}} \longrightarrow \boldsymbol{X}^{\boldsymbol{A}} \qquad (\text{resp.} \quad \boldsymbol{Y}^u f^{\boldsymbol{B}} = f^{\boldsymbol{A}} \boldsymbol{X}^u \colon \boldsymbol{X}^{\boldsymbol{B}} \longrightarrow \boldsymbol{Y}^{\boldsymbol{A}})$$

and hence canonically naturally isomorphic (37.1).

Similarly, given a deformable adjunction $f\colon \boldsymbol{X} \leftrightarrow \boldsymbol{Y} :g$ (43.1) and a functor $u\colon \boldsymbol{A} \to \boldsymbol{B}$, we will therefore say that f (resp. g) is **homotopically compatible** with homotopy u-colimit (resp. u-limit) functors if

(i)′ there exist u-colimit (resp. u-limit) and homotopy u-colimit (resp. u-limit) functors on $\boldsymbol{X}$ and $\boldsymbol{Y}$, and

(ii)′ the compositions (42.1) of a homotopy u-colimit (resp. u-limit) functor on $\boldsymbol{X}$ (resp. $\boldsymbol{Y}$) with a left (resp. a right) approximation of $f^{\boldsymbol{B}}$ (resp. $g^{\boldsymbol{B}}$) and of a left (resp. a right) approximation of $f^{\boldsymbol{A}}$ (resp. $g^{\boldsymbol{A}}$) with a homotopy u-colimit (resp. u-limit) functor on $\boldsymbol{Y}$ (resp. $\boldsymbol{X}$) are left (resp.

right) approximations of the canonically (naturally) isomorphic (37.1) compositions

$$X^A \xrightarrow{\text{colim}^u} X^B \xrightarrow{f^B} Y^B \quad \text{and} \quad X^A \xrightarrow{f^A} Y^A \xrightarrow{\text{colim}^u} Y^B$$

(resp. $\quad Y^A \xrightarrow{\lim^u} Y^B \xrightarrow{g^B} X^B \quad$ and $\quad Y^A \xrightarrow{g^A} X^A \xrightarrow{\lim^u} X^B \quad$)

respectively, and hence (38.4) are canonically weakly equivalent (37.5). Then one has the following

47.6. Sufficient conditions for homotopical compatibility of homotopy (co)limit functors with deformable adjoints. *Given a deformable adjunction $f \colon X \leftrightarrow Y \colon g$ (43.1) and a functor $u \colon A \to B$, a sufficient condition in order that f (resp. g) is homotopically compatible (47.5) with homotopy u-colimit (resp. u-limit) functors is that there exist u-colimit (resp. u-limit) functors on X and Y, and*

(i) *the pairs*

$$(\text{colim}^u \colon X^A \to X^B, f^B) \quad \text{and} \quad (f^A, \text{colim}^u \colon Y^A \to Y^B)$$

(resp. the pairs $\quad (\lim^u \colon Y^A \to Y^B, g^B) \quad$ *and* $\quad (g^A, \lim^u \colon X^A \to X^B) \quad$)

are left (resp. right) deformable (42.3)

which in particular is the case if

(ii) X *and* Y *are saturated (33.9) and the above pairs are locally left (resp. right) deformable (42.3).*

Proof. Part (i) is a consequence of 42.4, while part (ii) follows from 33.9(v) and 42.5 and the observation that it readily follows from the fact that the functors X^u and Y^u are homotopical and that g (resp. f) is right (resp. left) deformable (40.2), that the pairs

$$(X^u, g^B) \text{ and } (g^A, Y^u) \quad (\text{resp. } (Y^u, f^B) \text{ and } (f^A, X^u))$$

are right (resp. left) deformable (42.3).

48. Left and right systems

In order to be able to deal with the notions of *homotopical cocompleteness* and *completeness* (in §49), we first have to generalize the results of §42 and §44 on composable pairs of (not necessarily homotopical) functors between homotopical categories to diagram-like collections of such categories and functors which we will call *left* and *right systems*.

We thus start with defining these

48.1. Left and right systems. Given a category C, a **left** (resp. a **right**) C-**system** will be a function F which assigns

(i) to every object $D \in C$ a *homotopical category* FD
(ii) to every map $u \colon A \to B \in C$ a (not necessarily homotopical) *functor* $Fu \colon FA \to FB$, and
(iii) to every composable pair of maps $u \colon A \to B$ and $v \colon B \to D \in C$, a *natural weak equivalence* (called **composer**)

$$F(v, u) \colon (Fv)(Fu) \to F(vu) \quad (\text{resp. } F(v, u) \colon F(vu) \to (Fv)(Fu))$$

which is *associative* in the sense that, for every three composable maps $u\colon A \to B$, $v\colon B \to D$ and $x\colon D \to E \in \boldsymbol{C}$, the following diagram commutes

$$\begin{array}{ccc}
(Fx)(Fv)(Fu) \xrightarrow{(Fx)F(v,u)} (Fx)F(vu) & & (Fx)F(vu) \xrightarrow{(Fx)F(v,u)} (Fx)(Fv)(Fu) \\
{\scriptstyle F(x,v)(Fu)}\downarrow \quad\quad\quad \downarrow{\scriptstyle F(x,vu)} & (\text{resp.} \quad {\scriptstyle F(x,vu)}\uparrow \quad\quad\quad \uparrow{\scriptstyle F(x,v)(Fu)} &) \\
F(xv)(Fu) \xrightarrow[F(xv,u)]{} F(xvu) & & F(xvu) \xrightarrow[F(xv,u)]{} F(xv)(Fu)
\end{array}$$

and such a $\boldsymbol{C}$-system will be called

- (iv) **homotopical** if, for every map $u\colon A \to B \in \boldsymbol{C}$, the functor Fu is *homotopical*, and
- (v) **saturated** if, for every object $D \in \boldsymbol{C}$, FD is *saturated*, i.e. (33.9) a map in FD is a weak equivalence iff its image in $\operatorname{Ho} FD$ is an isomorphism.

Similarly there are

48.2. Maps and weak equivalences between systems. We only consider maps and weak equivalences between two left or right systems F and G when $FD = GD$ *for every object* $D \in \boldsymbol{C}$ and we then define a **map** $F \to G$ between two such left (resp. right) $\boldsymbol{C}$-systems F and G as a function h which assigns to every map $u\colon A \to B \in \boldsymbol{C}$ a natural transformation $hu\colon Fu \to Gu$, which is compatible with the composers in the sense that, for every composable pair of maps $u\colon A \to B$ and $v\colon B \to D \in \boldsymbol{C}$, the following diagram commutes

$$\begin{array}{ccc}
(Fv)(Fu) \xrightarrow{F(v,u)} F(vu) & & F(vu) \xrightarrow{F(v,u)} (Fv)(Fu) \\
{\scriptstyle (hv)(hu)}\downarrow \quad\quad \downarrow{\scriptstyle h(vu)} & (\text{resp.} \quad {\scriptstyle h(vu)}\downarrow \quad\quad \downarrow{\scriptstyle (hv)(hu)} &) \\
(Gv)(Gu) \xrightarrow[G(v,u)]{} G(vu) & & G(vu) \xrightarrow[G(v,u)]{} (Gv)(Gu)
\end{array}$$

and call such a map a **weak equivalence** whenever, for every map $u \in \boldsymbol{C}$, the natural transformation hu is a *natural weak equivalence*.

We denote the resulting homotopical category of left (resp. right) $\boldsymbol{C}$-systems by

$$\boldsymbol{C}_L\text{-}\mathbf{syst} \qquad (\text{resp. } \boldsymbol{C}_R\text{-}\mathbf{syst})$$

and its full subcategory spanned by the homotopical $\boldsymbol{C}$-systems (48.1) by

$$\boldsymbol{C}_L\text{-}\mathbf{syst}_{\mathrm{w}} \qquad (\text{resp. } \boldsymbol{C}_R\text{-}\mathbf{syst}_{\mathrm{w}}).$$

48.3. Example. If H is a $\boldsymbol{C}$-diagram of homotopical categories and (not necessarily homotopical) functors between them, then H gives rise to a left $\boldsymbol{C}$-system H_L and a right $\boldsymbol{C}$-system H_R which assigns to the objects and the maps of $\boldsymbol{C}$ the same homotopical categories and functors as H and in which, for every composable pair of maps $u\colon A \to B$ and $v\colon B \to D \in \boldsymbol{C}$, the composers (48.1) $H_L(v,u)$ and $H_R(v,u)$ are the *identity natural transformations* of the functor

$$H(vu) = (Hv)(Hu)\colon HA \to HD.$$

48.4. Example. For a *homotopical* left (resp. right) $\boldsymbol{C}$-system K (48.1) one can consider its **homotopy system**, i.e. the function $\operatorname{Ho} K$ which assigns (33.9)

(i) to every object $D \in \boldsymbol{C}$, the homotopy category $\operatorname{Ho} KD$, considered as a *minimal* homotopical category (33.6)

(ii) to every map $u\colon A \to B \in \boldsymbol{C}$, the induced functor $\operatorname{Ho} Ku\colon \operatorname{Ho} KA \to \operatorname{Ho} KB$, and

(iii) to every composable pair of maps $u\colon A \to B$ and $v\colon B \to D \in \boldsymbol{C}$, the induced natural isomorphism $\operatorname{Ho} K(v,u)$,

and note that (33.9)

(iv) $\operatorname{Ho} K$ *is a left (resp. a right)* $\boldsymbol{C}$*-system.*

Next we define

48.5. Left and right approximations of systems. Given a category $\boldsymbol{C}$ and a left or right $\boldsymbol{C}$-system F (48.1), the *left* or *right approximations* of F will, roughly speaking, be the *homotopical* $\boldsymbol{C}$-systems which

(i) consist of left or right approximations of the functors of F, and

(ii) in a homotopical sense, are closest to F from the left or the right.

More precisely, a **left** (resp. a **right**) **approximation** of a left (resp. a right) $\boldsymbol{C}$-system F will be a pair (K,a) consisting of an object (48.2)

$$K \in \boldsymbol{C}_L\text{-syst}_w \qquad (\text{resp. } K \in \boldsymbol{C}_R\text{-syst}_w)$$

and a map (48.2)

$$a\colon K \to F \in \boldsymbol{C}_L\text{-syst} \qquad (\text{resp. } a\colon F \to K \in \boldsymbol{C}_R\text{-syst})$$

such that

(iii) for every map $u \in \boldsymbol{C}$, the pair (Ku, au) is a *left* (resp. a *right*) approximation of Fu (41.1), and

(iv) (K,a) is a *homotopically terminal* (resp. *initial*) object of "the homotopical category of homotopical left (resp. right) $\boldsymbol{C}$-systems over (resp. under) F", i.e. the homotopical category

$$(\boldsymbol{C}_L\text{-syst}_w \downarrow F) \qquad (\text{resp. } (F \downarrow \boldsymbol{C}_R\text{-syst}_w))$$

which has as objects the above pairs (K,a) and, for every two such pairs (K_1, a_1) and (K_2, a_2) as maps and weak equivalences $(K_1, a_1) \to (K_2, a_2)$ the maps and the weak equivalences $t\colon K_1 \to K_2$ such that $a_2 t = a_1$ (resp. $t a_1 = a_2$).

It then follows readily from 37.6(vi), 38.3(i) and 48.2 that

(v) *if one object of* $(\boldsymbol{C}_L\text{-syst}_w \downarrow F)$ *(resp.* $(F \downarrow \boldsymbol{C}_R\text{-syst}_w))$ *satisfies* (iii) *and* (iv)*, then every object that satisfies one of* (iii) *and* (iv) *also satisfies the other*

so that (37.6(vi))

(vi) *the left (resp. right) approximations of F, if they exist, are homotopically unique* (37.5).

In order to obtain sufficient conditions for the existence of such left and right approximations we generalize the notions of *deformable* and *locally deformable composable* pairs of functors of 42.3 to those of

48.6. Deformable and locally deformable systems. Given a category C, we call a left (resp. a right) C-system F **locally left** (resp. **right**) **deformable** if

(i) there exists a function F_0 which associates with every object $D \in C$ a left (resp. a right) deformation retract $F_0 D \subset FD$ (40.1) such that, for every map $u\colon A \to B \in C$, the functor $Fu\colon FA \to FB$ is *homotopical* on $F_0 A$

which (40.2)(ii)′ is equivalent to requiring that

(i)′ there exists a **local left** (resp. a **right**) F-**deformation**, i.e. a function (r, s) which assigns to every object $D \in C$ a left (resp. a right) deformation (r_S, s_D) of FD (40.1) such that, for every map $u\colon A \to B \in C$, the functor $(Fu)r_A$ is *homotopical* and the natural transformation $(Fu)s_A r_A$ is a *natural weak equivalence*,

and we call F **left** (resp. **right**) **deformable** if

(ii) there exists a function F_0 as in (i) with the additional property that, for every map $u\colon A \to B \in C$, the functor $Fu\colon FA \to FB$ sends all of $F_0 A$ into $F_0 B$,

which (as in 42.3(v)) is equivalent to requiring that

(ii)′ there exists a **left** (resp. **right**) F-**deformation**, i.e. a local left (resp. right) F-deformation (r, s) (i)′ such that, for every composable pair of maps $u\colon A \to B$ and $v\colon B \to D \in C$, the natural transformation $(Fv)s_B(Fu)r_A$ is a *natural weak equivalence*

as for such a left (resp. right) F-deformation (r, s), the function F_0 which assigns to every object $D \in C$, the full subcategory $F_0 D \subset FD$ spanned by the union, taken over all objects $A \in C$ and finite sequences of maps

$$A \xrightarrow{u_1} \cdots \xrightarrow{u_n} D \quad \text{in } C \ (n \geq 0)$$

of the images of the functors

$$(Fu_n)\cdots(Fu_1)r_A\colon FA \longrightarrow FD \qquad (n \geq 0),$$

has the property that

(iii) for every map $v\colon D \to E \in C$, the functor $Fv\colon FD \to FE$ sends all of $F_0 D$ into $F_0 E$

and that

(iii)′ in view of 40.3 and the observation that the iterated composer (48.1)

$$(Fu_n)\cdots(Fu_1) \longrightarrow \cdots \longrightarrow F(u_n\cdots u_{k+1})(Fu_k)\cdots(Fu_1) \longrightarrow \cdots \longrightarrow F(u_n\cdots u_1)$$

is a natural weak equivalence, the functor $Fv\colon FD \to FE$ is homotopical on $F_0 D$.

Now we can state and prove the desired

48.7. Sufficient conditions for existence of approximations of systems. *Given a category C, a sufficient condition in order that a left (resp. a right) C-system F (48.1) has left (resp. right) approximations (48.5) is that F is left (resp. right) deformable (48.6(ii)).*

Proof (of the left half). Given a left $\boldsymbol{C}$-system F and a left F-deformation (r,s) (48.6(ii)$'$) denote, for every left $\boldsymbol{C}$-system K such that $KD = FD$ for every object $D \in \boldsymbol{C}$, by Kr the function which assigns

(i) to every object $D \in \boldsymbol{C}$, the homotopical category $(Kr)D = KD$,
(ii) to every map $u\colon A \to B \in \boldsymbol{C}$, the functor $(Kr)u = (Ku)r_A$, and
(iii) to every composable pair of maps $u\colon A \to B$ and $v\colon B \to D \in \boldsymbol{C}$, the composition $(Kr)(v,u)$ of the natural transformation

$$(Kr)v(Kr)u = (Kv)r_B(Ku)r_A \xrightarrow{(Kv)s_B(Ku)r_A} (Kv)(Ku)r_A$$

with the natural weak equivalence

$$(Kv)(Ku)r_A \xrightarrow{K(v,u)r_A} K(vu)r_A = (Kr)(vu)$$

and denote by Ks the function which assigns

(iv) to every map $u\colon A \to B \in \boldsymbol{C}$, the natural transformation

$$(Ks)u = (Ku)s_A.$$

A sufficient condition in order that Kr so defined is a left $\boldsymbol{C}$-system is that, for every composable pair of maps $u\colon A \to B$ and $v\colon B \to D \in \boldsymbol{C}$, the natural transformation $(Kv)s_B(Ku)r_A$ is a natural weak equivalence and it follows that

(v) if K is a homotopical left $\boldsymbol{C}$-system (48.1(iv)), then so is Kr and the function Ks is a weak equivalence $Kr \to K \in \boldsymbol{C}_L\text{-syst}_\mathrm{w}$ (48.5), and
(vi) Fr is a homotopical left $\boldsymbol{C}$-system and the function Fs is a map $Fr \to F \in \boldsymbol{C}_L\text{-syst}$ (48.5),

and that consequently, in view of the *naturality* in K of the functions Kr and Ks,

(vii) every object $(K,a) \in (\boldsymbol{C}_L\text{-syst}_\mathrm{w} \downarrow F)$ (48.5) gives rise to a zigzag in $(\boldsymbol{C}_L\text{-syst}_\mathrm{w} \downarrow F)$

$$(K,a) \xleftarrow{Ks} (Kr, as) \xrightarrow{ar} (Fr, Fs)$$

in which as denotes the diagonal of the commutative diagram

$$\begin{array}{ccc} Kr & \xrightarrow{ar} & Fr \\ {\scriptstyle Ks}\downarrow & & \downarrow{\scriptstyle Fs} \\ K & \xrightarrow{a} & F \end{array}$$

which is natural in (K,a) and in which $ar\colon Kr \to Fr$ is the map given by $(ar)u = (au)r_A$ for every map $u\colon A \to B \in \boldsymbol{C}$,

The desired result now follows from 38.2 and 48.6(i)$'$ and the observation that, as for every map $u\colon A \to B \in \boldsymbol{C}$, the natural transformation

$$((Fs)r)u = (Fs)ur_A = (Fu)s_A r_A$$

is a natural weak equivalence, the map $(Fs)r\colon (Fr)r \to Fr \in \boldsymbol{C}_L\text{-syst}_\mathrm{w}$ is a weak equivalence.

It remains to generalize to systems the rather useful sufficient condition (42.5) in order that a composable pair of functors between saturated homotopical categories be deformable (42.3), namely that the pair is locally deformable (42.3) and has a deformable pair of adjoints.

To do this we thus first have to define

48.8. Adjunctions between left and right systems. Given a category C, a left C-system F and a right C^{op}-system F', F is said to be a **left adjoint** of F' and F' is said to be a **right adjoint** of F, if $FD = F'D$ for every object $D \in C$ and there is *given* an **adjunction** $F \leftrightarrow F'$, i.e. there are given adjunctions

$$Fu\colon FA \longleftrightarrow F'B : F'u^{\mathrm{op}},$$

one for every map $u\colon A \to B \in C$, such that, for every composable pair of maps $u\colon A \to B$ and $v\colon B \to D \in C$, the composers $F(v, u)$ and $F'(u^{\mathrm{op}}, v^{\mathrm{op}})$ are conjugate (39.10) with respect to the adjunction

$$F(vu)\colon FA \longleftrightarrow F'D : F'(u^{\mathrm{op}} v^{\mathrm{op}})$$

and the composition of the adjunctions

$$Fu\colon FA \longleftrightarrow F'B : F'u^{\mathrm{op}} \quad \text{and} \quad Fv\colon FB \longleftrightarrow F'D : F'v^{\mathrm{op}}\ .$$

Moreover the functions ε and η which assign to every map $u\colon A \to B \in C$ the counit and the unit

$$\varepsilon u\colon (Fu)(F'u^{\mathrm{op}}) \longrightarrow 1_{FB} \quad \text{and} \quad \eta u^{\mathrm{op}}\colon 1_{FA} \longrightarrow (F'u^{\mathrm{op}})(Fu)$$

of the adjunction $Fu\colon FA \leftrightarrow FB : F'u^{\mathrm{op}}$, will be called the **counit** and the **unit** of the adjunction $F \leftrightarrow F'$.

One then readily verifies that (39.10)

(i) *if a left (resp. a right) C-system has a right (resp. a left) adjoint, then all such adjoints are canonically isomorphic (37.1), and*

(ii) *if F is a left (resp. a right) C-system of which the composers are natural isomorphisms, then F has a right (resp. a left) adjoint iff, for every map $u\colon A \to B \in C$, the functor $Fu\colon FA \to FB$ has a right (resp. a left) adjoint.*

We also define

48.9. Deformable and locally deformable adjunctions of systems. Given a category C, an adjunction $F \leftrightarrow F'$ (48.8) between a left C-system F and a right C^{op}-system F' will be called

(i) **locally deformable**, if F and F' are locally left and right deformable (48.6(i)) respectively, and

(ii) **deformable** if F and F' are left and right deformable (48.6(ii)).

Now we can state the above mentioned

48.10. Sufficient conditions for deformability of systems. *Let C be a category and $F \leftrightarrow F'$ a locally deformable adjunction (48.8 and 48.9) between a saturated left C-system F (48.1(v)) and a saturated right C^{op}-system F'. Then F is left deformable (48.8(ii)) iff F' is right deformable.*

Proof (of the "if" part). If (48.6) (r,s) and (r',s') are a local left F-deformation and a right F'-deformation then, for every composable pair of maps $u\colon A \to B$ and $v\colon G \to D \in \boldsymbol{C}$, the natural transformation $(F'u^{\mathrm{op}})s'_B(F'v^{\mathrm{op}})r'_D$ is a natural weak equivalence and hence, in view of the saturation of FD and 42.6, so is the natural transformation $(Fv)s_B(Fu)r_A$, which implies that (r,s) is actually a left F-deformation.

We end with noting the existence of

48.11. Derived adjunctions of deformable adjunctions of systems. *Let $\boldsymbol{C}$ be a category and let $F \leftrightarrow F'$ be a deformable adjunction (48.8 and 48.9) between a left $\boldsymbol{C}$-system F (48.1) and a right $\boldsymbol{C}$-system F' Then the adjunction $F \leftrightarrow F'$ induces, for every left approximation (K,a) of F and every right approximation (K',a') of F' (48.5) (48.5 and 48.7) a **derived adjunction** (48.4)*

$$\mathrm{Ho}\, K \longleftrightarrow \mathrm{Ho}\, K'$$

which (48.8(ii)), for every map $u\colon A \to B \in \boldsymbol{C}$, consists of the derived adjunction (44.2)

$$\mathrm{Ho}\, Ku\colon \mathrm{Ho}\, FA \longleftrightarrow \mathrm{Ho}\, F'B \colon \mathrm{Ho}\, K'u^{\mathrm{op}}$$

associated with the left approximation (Ku, au) of Fu and the right approximation $(K'u^{\mathrm{op}}, a'u^{\mathrm{op}})$ of $F'u^{\mathrm{op}}$.

Proof. It follows from 44.4 that, for every composable pair of maps $u\colon A \to B$ and $v\colon B \to D \in \boldsymbol{C}$, the composition

$$\mathrm{Ho}(Kv)(Ku)\colon \mathrm{Ho}\, \boldsymbol{X}^A \longleftrightarrow \mathrm{Ho}\, \boldsymbol{X}^D \colon \mathrm{Ho}(K'u^{\mathrm{op}})(K'v^{\mathrm{op}})$$

of the derived adjunctions (44.2)

$$\mathrm{Ho}\, Ku\colon \mathrm{Ho}\, \boldsymbol{X}^A \longleftrightarrow \mathrm{Ho}\, \boldsymbol{X}^B \colon \mathrm{Ho}\, K'u^{\mathrm{op}}$$

and $\quad \mathrm{Ho}\, Kv\colon \mathrm{Ho}\, \boldsymbol{X}^B \longleftrightarrow \mathrm{Ho}\, \boldsymbol{X}^D \colon \mathrm{Ho}\, K'v^{\mathrm{op}}$

associated with the approximations

$$Ku \xrightarrow{au} Fu, \quad F'u^{\mathrm{op}} \xrightarrow{a'u^{\mathrm{op}}} K'u^{\mathrm{op}}, \quad Kv \xrightarrow{av} Fv \quad \text{and} \quad F'v^{\mathrm{op}} \xrightarrow{a'v^{\mathrm{op}}} K'v^{\mathrm{op}}$$

is the derived adjunction associated with the approximations (42.4)

$$(Kv)(Ku) \xrightarrow{(av)(au)} (Fv)(Fu) \quad \text{and} \quad (F'u^{\mathrm{op}})(F'v^{\mathrm{op}}) \xrightarrow{(a'u^{\mathrm{op}})(a'v^{\mathrm{op}})} (K'u^{\mathrm{op}})(K'v^{\mathrm{op}})$$

We then have to show that the natural isomorphisms

$$\mathrm{Ho}\, K(v,u) \quad \text{and} \quad \mathrm{Ho}\, K'(u^{\mathrm{op}}, v^{\mathrm{op}})$$

are conjugate with respect to this composite derived adjunction and the derived adjunction

$$\mathrm{Ho}\, K(vu)\colon \mathrm{Ho}\, \boldsymbol{X}^A \longleftrightarrow \mathrm{Ho}\, \boldsymbol{X}^D \colon \mathrm{Ho}\, K'(u^{\mathrm{op}} v^{\mathrm{op}})$$

associated with the approximations

$$K(vu) \xrightarrow{a(vu)} F(vu) \quad \text{and} \quad F'(u^{\mathrm{op}} v^{\mathrm{op}}) \xrightarrow{a'(u^{\mathrm{op}} v^{\mathrm{op}})} K'(u^{\mathrm{op}} v^{\mathrm{op}})$$

and do this by noting that this follows from 44.3 and the observation that in view of 41.6(vi) the commutative diagrams

$$\begin{array}{ccc} (Kv)(Ku) & \xrightarrow{(av)(au)} & (Fv)(Fu) \\ {\scriptstyle K(v,u)}\downarrow & & \downarrow{\scriptstyle F(vu)} \\ K(vu) & \xrightarrow{a(vu)} & F(vu) \end{array}$$

and

$$\begin{array}{ccc} (F'u^{\mathrm{op}})(F'v^{\mathrm{op}}) & \xrightarrow{(a'u^{\mathrm{op}})(a'v^{\mathrm{op}})} & (K'u^{\mathrm{op}})(K'v^{\mathrm{op}}) \\ {\scriptstyle F'(u^{\mathrm{op}},v^{\mathrm{op}})}\uparrow & & \uparrow{\scriptstyle K'(u^{\mathrm{op}},v^{\mathrm{op}})} \\ F'(u^{\mathrm{op}}v^{\mathrm{op}}) & \xrightarrow{a'(u^{\mathrm{op}}v^{\mathrm{op}})} & K'(u^{\mathrm{op}}v^{\mathrm{op}}) \end{array}$$

are left and right approximations respectively of the deformable conjugate pair of natural transformations (43.1)

$$F(v, u) \quad \text{and} \quad F(u^{\mathrm{op}}, v^{\mathrm{op}}) \ .$$

49. Homotopical cocompleteness and completeness (special case)

We now use the results of the preceding two sections to obtain, for homotopical categories which are cocomplete and complete, notions of homotopical cocompleteness and completeness which are considerably stronger than requiring the existence of homotopy D-colimit or D-limit functors for every small category D, or even of homotopy u-colimit or u-limit functors for every functor $u\colon A \to B$ between small categories.

We start with a brief discussion of the categorical notions of

49.1. Cocompleteness and completeness. Let **cat** denote the category of the *small categories* (32.3) and for every homotopical category X, let $X^{(\mathbf{cat})}$ denote the $\mathbf{cat}^{\mathrm{op}}$-diagram which associates with every object $D \in \mathbf{cat}$ the diagram category X^D and with every map $u\colon A \to B \in \mathbf{cat}$ the induced diagram functor $X^u\colon X^B \to X^A$ (47.1(i)′). Then X is called **cocomplete** (resp. **complete**) if, for every object $D \in \mathbf{cat}$, there exists a D-colimit (resp. D-limit) functor on X (47.1(i)) and such cocompleteness (resp. completeness) implies that

(i) *for every map $u\colon A \to B \in \mathbf{cat}$ there exist u-colimit (resp. u-limit) functors on X (for instance the functor which sends an object $T \in X^A$ to the functor $B \to X$ which associates with each object $B \in B$ the object (47.1(ii)′)*

$$\mathrm{colim}^{(u\downarrow B)} j^*T \qquad (\text{resp. } \lim{}^{(B\downarrow u)} j^*T)$$

where

$$j\colon (u\downarrow B) \longrightarrow A \qquad (\text{resp. } j\colon (B\downarrow u) \longrightarrow A)$$

denotes the forgetful functor), and

(ii) *for every composable pair of maps $u\colon A \to B$ and $v\colon B \to D \in \mathbf{cat}$, every composition of a u-colimit (resp. u-limit) functor on X with a v-colimit (resp. v-limit) functor is a vu-colimit (resp. vu-limit) functor,*

However even though in general it is not possible to choose for every map $u \in \mathbf{cat}$ a u-colimit (resp. u-limit) functor on $\mathbf{X}$ such that these functors form a *diagram* indexed by **cat**,

  (iii) there exists what we will call a **colimit**(resp. a **limit**) **system** on $\mathbf{X}$, i.e. a left (resp. a right) adjoint of the right (resp. left) $\mathbf{cat}^{\mathrm{op}}$-system $\mathbf{X}_R^{(\mathbf{cat})}$ (resp. $\mathbf{X}_L^{(\mathbf{cat})}$) (48.3 and 48.8).

Clearly (48.8(i) and (ii))

  (iv) *such a colimit (resp. limit) system on a category* $\mathbf{X}$, *if it exists, is categorically unique* (37.1), *and*
  (v) *such a colimit (resp. limit) system on a category* $\mathbf{X}$ *exists iff* $\mathbf{X}$ *is cocomplete (resp. complete)*.

Moreover

  (vi) *if* $\mathbf{X}$ *is cocomplete (resp. complete), then the function*

$$\mathrm{colim}^{(\mathbf{cat})} \qquad (resp.\ \mathrm{lim}^{(\mathbf{cat})})$$

  given by the formulas (47.1 and 47.3)

$$\mathrm{colim}^{(\mathbf{cat})} u = \mathrm{colim}^u \qquad (resp.\ \mathrm{lim}^{(\mathbf{cat})} u = \mathrm{lim}^u)$$

  and

$$\mathrm{colim}^{(\mathbf{cat})}(v, u) = \mathrm{colim}^{(v,u)} \qquad (resp.\ \mathrm{lim}^{(\mathbf{cat})}(v, u) = \mathrm{lim}^{(v,u)})$$

  is such a colimit (resp. limit) system on $\mathbf{X}$.

Now we turn to

49.2. Homotopical cocompleteness and completeness. If $\mathbf{X}$ is a homotopical category, then

  (i) the existence of homotopy $\mathbf{D}$-colimit (or $\mathbf{D}$-limit) functors on $\mathbf{X}$ for every small category $\mathbf{D}$, i.e. object $\mathbf{D} \in \mathbf{cat}$ (32.3)

need not imply

  (ii) the existence of homotopy u-colimit (or a u-limit) functors on $\mathbf{X}$ for every map $u \in \mathbf{cat}$,

nor need (ii) imply that

  (iii) for every composable pair of maps $u\colon \mathbf{A} \to \mathbf{B}$ and $v\colon \mathbf{B} \to \mathbf{D} \in \mathbf{cat}$, the compositions of homotopy u-colimit (or u-limit) functors on $\mathbf{X}$ with homotopy v-colimit (or v-limit) functors on $\mathbf{X}$ are homotopy vu-colimit (or vu-limit) functors on $\mathbf{X}$.

In view of 49.1(v) we therefore will call a homotopical category $\mathbf{X}$ **homotopically cocomplete** (resp. **complete**) if

  (iv) $\mathbf{X}$ is *cocomplete* (resp. *complete*) and
  (v) there exists a **homotopy colimit** (resp. **limit**) **system** on $\mathbf{X}$,

which we will define as a *left* (resp. a *right*) *approximation* (48.5) of the colimit (resp. the limit) system $\mathrm{colim}^{(\mathbf{cat})}$ (resp. $\mathrm{lim}^{(\mathbf{cat})}$).

Clearly (48.5(vi))

  (vi) *such homotopy colimit (resp. limit) systems on* $\mathbf{X}$, *if they exist, are homotopically unique* (37.5).

Moreover one has, in view of 48.6, 48.7 and 48.10, the following sufficient conditions for their existence, i.e.

49.3. Sufficient conditions for homotopical cocompleteness and completeness. *Sufficient conditions for the homotopical cocompleteness (resp. completeness) of a homotopical category X are that*

(i) X *is cocomplete (resp. complete) and the colimit (resp. the limit) system* $\operatorname{colim}^{(\mathbf{cat})}$ *(resp.* $\lim^{(\mathbf{cat})}$*) on X is left (resp. right) deformable, i.e. there exists a function F_0 which assigns to every object $D \in \mathbf{cat}$ a left (resp. a right) deformation retract $F_0 D \subset X^D$ (40.1) such that, for every map $u: A \to B \in \mathbf{cat}$, the functor $\operatorname{colim}^u: X^A \to X^B$ (resp. $\lim^u: X^A \to X^B$*)

(i)′ *is homotopical on $F_0 A$, and*
(i)″ *sends all of $F_0 A$ into $F_0 B$*

which, in view of the right (resp. the left) deformability of $X_R^{(\mathbf{cat})}$ (resp. $X_L^{(\mathbf{cat})}$) (48.3 and 49.1), is in particular the case if

(ii) X *is cocomplete (resp. complete) and saturated (33.9 and 48.1(v)) and the colimit (resp. the limit) system* $\operatorname{colim}^{(\mathbf{cat})}$ *(resp.* $\lim^{(\mathbf{cat})}$*) on X is locally left (resp. right) deformable, i.e. there exists a function F_0 which assigns to every object $D \in \mathbf{cat}$ a left (resp. a right) deformation retract $F_0 D \subset X^D$ such that, for every map $u: A \to B \in \mathbf{cat}$, the functor $\operatorname{colim}^u: X^A \to X^B$ (resp. $\lim^u: X^A \to X^B$) satisfies (i)′ but not necessarily (i)″ (in which case there exists a possibly different such function which satisfies both (i)′ and (i)″).*

We end with noting that 48.11 implies that, under the assumptions made in 49.3, the existence of

49.4. Derived adjunctions of homotopy colimit and limit systems. *Let X be a homotopical category which satisfies 49.3(i). Then every homotopy colimit (resp. limit) system (K, a) on X comes with the derived adjunction*

$$\operatorname{Ho} K \longleftrightarrow \operatorname{Ho} X_R^{(\mathbf{cat})} \qquad (\text{resp. } \operatorname{Ho} X_L^{(\mathbf{cat})} \longleftrightarrow \operatorname{Ho} K)$$

associated with (K, a) and the pair consisting of $X_R^{(\mathbf{cat})}$ (resp. $X_L^{(\mathbf{cat})}$) and its identity map.

50. Homotopical colimit and limit functors

The *homotopy* colimit and limit functors which we considered in the preceding sections assume the *existence* of colimit or limit functors and *choice* of an arbitrary but fixed such functor, and even though it seems, for the moment at least, that this is the only situation in which one is really interested, we will now discuss a notion of what we will call *homotopical* colimit and limit functors which in the following sense "generalize" the homotopy colimit and limit functors:

Given a homotopical category X and a functor $u: A \to B$, we define a *homotopical u-colimit* (or *u-limit*) *functor on X* as a homotopically terminal (or initial) Kan extension (38.6) of the identity functor along the induced diagram functor $X^u: X^B \to X^A$, and note that as [**Mac71**, Ch. X, §7]

(i) the left (or right) adjoints of $\boldsymbol{X}^u$ (i.e. the u-colimit (or u-limit) functors on $\boldsymbol{X}$) are exactly the same as the terminal (or initial) Kan extensions of the identity functor of $\boldsymbol{X}^B$ along $\boldsymbol{X}^u$,

and as

(ii) for every terminal (or initial) Kan extension f of the identity functor of $\boldsymbol{X}^B$ along $\boldsymbol{X}^u$, the left (or right) approximations of f (41.1) are the f-*presentations* of the homotopically terminal (or initial) Kan extensions of the identity functor along $\boldsymbol{X}^u$, in the sense that they are the approximations associated with these homotopical Kan extensions under a 1-1 correspondence which is induced by the counit (or unit) of the Kan extension f,

it follows that

(iii) for a cocomplete (or complete) $\boldsymbol{X}$, the *homotopy u-colimit* (or *u-limit*) functors on $\boldsymbol{X}$ are exactly the colim^u- (or lim^u-)*presentations* of the *homotopical u-colimit* (or *u-limit*) functors on $\boldsymbol{X}$.

To do this we first have to discuss

50.1. Kan extensions of the identity and sufficient conditions for their existence. Given a functor $g\colon \boldsymbol{Y} \to \boldsymbol{X}$, we recall from 38.6 that a **terminal** (resp. an **initial**) **Kan extension of the identity** (i.e. $1_{\boldsymbol{Y}}$) along g is a functor $f\colon \boldsymbol{X} \to \boldsymbol{Y}$, together with an often suppressed natural transformation

$$e\colon fg \longrightarrow 1_{\boldsymbol{Y}} \qquad (\text{resp. } e\colon 1_{\boldsymbol{Y}} \longrightarrow fg),$$

called **counit** (resp. **unit**) such that the pair (f, e) is a terminal (resp. an initial) object of the category (38.6)

$$(-g \downarrow 1_{\boldsymbol{Y}}) \qquad (\text{resp. } (1_{\boldsymbol{Y}} \downarrow -g)) \ .$$

Thus (37.2(vi))

(i) *such Kan extensions of the identity, if they exist, are categorically unique* (37.1).

Moreover [**Mac71**, Ch. X, §7]

(ii) *for every functor $g\colon \boldsymbol{Y} \to \boldsymbol{X}$ which has left (resp. right) adjoints (46.4), the pairs (f, e) consisting of a terminal (resp. an initial) Kan extension of the identity along g and its counit (resp. unit) are exactly the same as the pairs (f, e) consisting of a left (resp. a right) adjoint of g and the counit (resp. the unit) of its adjunction*

and hence

(iii) *given a functor $g\colon \boldsymbol{Y} \to \boldsymbol{X}$, a sufficient condition for the existence of a terminal (resp. an initial) Kan extension of the identity along g is the existence of a left (resp. a right) adjoint of g.*

Next we consider

50.2. Homotopical Kan extensions of the identity and sufficient conditions for their existence. Given a (not necessarily homotopical) functor $g\colon Y \to X$ between homotopical categories, we recall from 38.6 that a **homotopically terminal** (resp. **initial**) **Kan extension of the identity** (i.e. 1_Y) along g is a *homotopical* functor $k\colon X \to Y$, together with an often suppressed natural transformation
$$d\colon kg \longrightarrow 1_Y \qquad (\text{resp. } d\colon 1_Y \longrightarrow kg)$$
such that the pair (k, d) is a homotopically terminal (resp. initial) object (38.2) of the category (38.6)
$$(-g \downarrow 1_Y)_w \qquad (\text{resp. } (1_Y \downarrow -g)_w)\ .$$
Thus (37.6(vi))

(i) *such homotopical Kan extensions of the identity, if they exist, are homotopically unique (37.5).*

Moreover one readily verifies that

(ii) *for every (not necessarily homotopical) functor $g\colon Y \to X$ between homotopical categories, every terminal (resp. initial) Kan extension f of the identity along g gives rise to a 1-1 correspondence between the left (resp. the right) approximations of f (41.1) and the homotopically terminal (resp. initial) Kan extensions of the identity along g, which is induced by the isomorphism (41.1)*
$$\big(\mathrm{Fun}_w(X, Y) \downarrow f\big) \approx (-g \downarrow 1_Y)_w \qquad \big(\text{resp. } \big(f \downarrow \mathrm{Fun}_w(X, Y)\big) \approx (1_Y \downarrow -g)_w\big)$$
which sends an object (k, a) to the pair (k, d) in which, if e denotes the counit (resp. the unit) of f (50.1), the natural transformation d is the composition
$$kg \xrightarrow{ag} fg \xrightarrow{e} 1_Y \qquad (\text{resp. } 1_Y \xrightarrow{e} fg \xrightarrow{ag} kg).$$

This together with 41.2 implies that

(iii) *given a functor $g\colon Y \to X$, a sufficient condition for the existence of a homotopically terminal (resp. initial) Kan extension of the identity along g is the existence of a left (resp. right) deformable (40.2) terminal (resp. initial) Kan extension of the identity along g.*

We also note that, in view of (ii), it is convenient

(iv) *given a functor $g\colon Y \to X$, a terminal (resp. initial) Kan extension f of the identity along g, and a pair consisting of a left (resp. a right) approximation (k, a) of f (41.1) and the corresponding (as in (ii)) homotopically terminal (resp. initial) Kan extension (k, d) of the identity along g, to say that (k, a) is the f-**presentation** of (k, d).*

Next we discuss

50.3. Sufficient conditions for composability of Kan extensions and homotopical Kan extensions. For every composable pair of (not necessarily homotopical) functors
$$g_2\colon Z \longrightarrow Y \qquad \text{and} \qquad g_1\colon Y \longrightarrow X$$

between homotopical categories one can consider the *homotopical* **composition functors**

$$(-g_1 \downarrow 1_{\boldsymbol{Y}}) \times (-g_2 \downarrow 1_{\boldsymbol{Z}}) \longrightarrow (-g_1g_2 \downarrow 1_{\boldsymbol{Z}})$$
$$(\text{resp. } (1_{\boldsymbol{Y}} \downarrow -g_1) \times (1_{\boldsymbol{Z}} \downarrow -g_2) \longrightarrow (1_{\boldsymbol{Z}} \downarrow -g_1g_2))$$

and

$$(-g_1 \downarrow 1_{\boldsymbol{Y}})_{\mathrm{w}} \times (-g_2 \downarrow 1_{\boldsymbol{Z}})_{\mathrm{w}} \longrightarrow (-g_1g_2 \downarrow 1_{\boldsymbol{Z}})_{\mathrm{w}}$$
$$(\text{resp. } (1_{\boldsymbol{Y}} \downarrow -g_1)_{\mathrm{w}} \times (1_{\boldsymbol{Z}} \downarrow -g_2)_{\mathrm{w}} \longrightarrow (1_{\boldsymbol{Z}} \downarrow -g_1g_2)_{\mathrm{w}})$$

which send an object $((f_1, e_1), (f_2, e_2))$ to the pair (f_2f_1, e_2e_1) (resp. (f_2f_1, e_1e_2)) where e_2e_1 (resp. e_1e_2) denotes the composite natural transformation

$$f_2f_1g_1g_2 \xrightarrow{e_1} f_2g_2 \xrightarrow{e_2} 1_{\boldsymbol{Z}} \quad (\text{resp. } 1_{\boldsymbol{Z}} \xrightarrow{e_2} f_2g_2 \xrightarrow{e_1} f_2f_1g_1g_2).$$

It then follows from 39.10(iii), 46.4 and 50.1(ii) that

(i) *a sufficient condition in order that, for two terminal (resp. initial) Kan extensions of the identity f_1 and f_2 along g_1 and g_2 respectively, their composition f_2f_1 is similar Kan extension along g_1g_2, is that g_1 and g_2 both have left (resp. right) adjoints, in which case (50.1(ii)) f_1 and f_2 are such adjoints and their composition as Kan extensions coincides with their composition as adjoints (39.10(iii)).*

Furthermore, it follows from 50.2(ii) that

(ii) *for every two terminal (resp. initial) Kan extensions of the identity f_1 and f_2 along g_1 and g_2 respectively for which the composition f_2f_1 is a similar Kan extension along g_1g_2, the diagram*

$$\begin{array}{ccc}
\big(\mathrm{Fun}_{\mathrm{w}}(\boldsymbol{X},\boldsymbol{Y}) \downarrow f_1\big) \times \big(\mathrm{Fun}_{\mathrm{w}}(\boldsymbol{Y},\boldsymbol{Z}) \downarrow f_2\big) & \longrightarrow & \big(\mathrm{Fun}_{\mathrm{w}}(\boldsymbol{X},\boldsymbol{Z}) \downarrow f_2f_1\big) \\
\approx \downarrow & & \downarrow \approx \\
(-g_1 \downarrow 1_{\boldsymbol{Y}})_{\mathrm{w}} \times (-g_2 \downarrow 1_{\boldsymbol{Z}})_{\mathrm{w}} & \longrightarrow & (-g_1g_2 \downarrow 1_{\boldsymbol{Z}})_{\mathrm{w}}
\end{array}$$

(resp.
$$\begin{array}{ccc}
\big(f_1 \downarrow \mathrm{Fun}_{\mathrm{w}}(\boldsymbol{X},\boldsymbol{Y})\big) \times \big(f_2 \downarrow \mathrm{Fun}_{\mathrm{w}}(\boldsymbol{Y},\boldsymbol{Z})\big) & \longrightarrow & \big(f_2f_1 \downarrow \mathrm{Fun}_{\mathrm{w}}(\boldsymbol{X},\boldsymbol{Z})\big) \\
\approx \downarrow & & \downarrow \approx \\
(1_{\boldsymbol{Y}} \downarrow -g_1)_{\mathrm{w}} \times (1_{\boldsymbol{Z}} \downarrow -g_2)_{\mathrm{w}} & \longrightarrow & (1_{\boldsymbol{Z}} \downarrow -g_1g_2)_{\mathrm{w}}
\end{array})$$

in which the horizontal maps are the composition functors and the vertical ones are as in 50.2(ii), commutes.

and this, together with 42.4, implies that

(iii) *if f_1 and f_2 are terminal (resp. initial) Kan extensions of the identity along g_1 and g_2 respectively for which the composition f_2f_1 is a similar Kan extension along g_1g_2, then a sufficient condition in order that every composition of a homotopically terminal (resp. initial) Kan extension of the identity along g_1 with one along g_2 is one along g_2g_2 is that*

(iii)' *the pair (f_1, f_2) is left (resp. right) deformable (42.3)*

Now we are ready to deal with

50.4. Homotopical colimit and limit functors. Given a homotopical category $\boldsymbol{X}$ and a functor $u\colon \boldsymbol{A} \to \boldsymbol{B}$, a **homotopical u-colimit** (resp. **u-limit**) **functor** on $\boldsymbol{X}$ will be a homotopically terminal (resp. initial) Kan extension of the identity (50.2) along the induced diagram functor $\boldsymbol{X}^u \colon \boldsymbol{X}^{\boldsymbol{B}} \to \boldsymbol{X}^{\boldsymbol{A}}$ (47.1). It then follows from 50.2(i), 50.2(iii) and 50.3(iii) that

(i) *such homotopical colimit and limit functors, if they exist, are homotopically unique* (37.5),

(ii) *a sufficient condition for the existence of such a homotopical u-colimit (resp. u-limit) functor on $\boldsymbol{X}$ is the existence of a left (resp. a right) deformable (40.2) terminal (resp. initial) Kan extension of the identity along the induced diagram functor $\boldsymbol{X}^u$, and*

(iii) *given two composable functors $u\colon \boldsymbol{A} \to \boldsymbol{B}$ and $v\colon \boldsymbol{B} \to \boldsymbol{D}$, a sufficient condition in order that every composition of a homotopical u-colimit (resp. u-limit) functor on $\boldsymbol{X}$ with a homotopical v-colimit (resp. v-limit) functor on $\boldsymbol{X}$ is a homotopical vu-colimit (resp. vu-limit) functor on $\boldsymbol{X}$ is that*

(iii)′ *there exist terminal (resp. initial) Kan extensions of the identity f_1 and f_2 along $\boldsymbol{X}^u$ and $\boldsymbol{X}^v$ respectively for which the composition $f_2 f_1$ is a similar Kan extension along $\boldsymbol{X}^{vu}$, and*

(iii)″ *the pair (f_1, f_2) is left (resp. right) deformable (42.3).*

We end with two propositions which formalize the statement that, in the case where there exist colimit and limit functors, the above *homotopical* colimit and limit functors are "essentially the same" as the *homotopy* colimit and limit functors which we considered in §47.

50.5. Relation between homotopy (co)limit functors and homotopical ones. *Let $\boldsymbol{X}$ be a homotopical category and let $u\colon \boldsymbol{A} \to \boldsymbol{B}$ be a functor for which there exist u-colimit (resp. u-limit) functors on $\boldsymbol{X}$. Then*

(i) *there exist homotopical u-colimit (resp. u-limit) functors on $\boldsymbol{X}$* (50.4)

iff

(ii) *there exist homotopy u-colimit (resp. u-limit) functors on $\boldsymbol{X}$* (47.1)

and

(iii) *if (i) and (ii) hold, then the homotopy u-colimit (resp. u-limit) functors on $\boldsymbol{X}$ are exactly the colim^u- (resp. $\lim^u$-) presentations (50.2(iv)) of the homotopical u-colimit (resp. u-limit) functors on $\boldsymbol{X}$.*

Proof. This follows from 50.1(iii) and 50.2(ii).

50.6. Relation between compositions of homotopy (co)limit functors and of homotopical ones. *Let $\boldsymbol{X}$ be a homotopical category and let $u\colon \boldsymbol{A} \to \boldsymbol{B}$ and $v\colon \boldsymbol{B} \to \boldsymbol{D}$ be functors for which there exist u- and v-colimit (resp. u- and v-limit) functors on $\boldsymbol{X}$. Then*

(i) *every composition of a homotopical u-colimit (resp. u-limit) functor on $\boldsymbol{X}$ with a homotopical v-colimit (resp. v-limit) functor on $\boldsymbol{X}$ is a homotopical vu-colimit (resp. vu-limit) functor on $\boldsymbol{X}$*

iff

and
(ii) *every "composition" (47.3) of a homotopy u-colimit (resp. u-limit) functor on X (47.1) with a homotopy v-colimit (resp. v-limit) functor on X is a homotopy vu-colimit (resp. vu-limit) functor on X*

(iii) *if (i) and (ii) hold, then, for every pair consisting of a homotopical u-colimit (resp. u-limit) functor (k_u, d_u) on X with (k_u, a_u) as its colim^u- (resp. $\lim^u$-) presentation (50.5) and homotopical v-colimit (resp. v-limit) functor (k_v, d_v) on X with (k_v, a_v) as its colim^v- (resp. $\lim^v$-) presentation, the "composition" (47.3) of (k_u, a_u) with (k_v, a_v) is the $\operatorname{colim}^{vu}$- (resp. $\lim^{vu}$-) presentation (50.2(iv)) of the composition of (k_u, d_u) and (k_v, d_v) (50.3).*

Proof (of the colimit half). This follows from 50.3(ii) and the commutativity of the diagram

$$\begin{array}{ccc} (\operatorname{Fun}_w(X^A, X^D) \downarrow \operatorname{colim}^v \operatorname{colim}^u) & \xrightarrow{\operatorname{colim}_*^{(v,u)}}_{\approx} & (\operatorname{Fun}_w(X^A, X^D) \downarrow \operatorname{colim}^{vu}) \\ \approx \downarrow & & \downarrow \approx \\ (-X^u X^v \downarrow 1_{X^D})_w & \xrightarrow{1} & (-X^{vu} \downarrow 1_{X^D})_w \end{array}$$

in which the vertical maps are as in 50.2(ii).

51. Homotopical cocompleteness and completeness (general case)

In this last section we describe, for arbitrary homotopical categories, notions of homotopical cocompleteness and completeness which, for categories which are cocomplete or complete, reduce to the one we considered in §49, by calling a homotopical category X *homotopically cocomplete* if there exist what we will call *homotopical colimit systems* on X which we define as "homotopically terminal Kan extensions along the right $\mathbf{cat}^{\mathrm{op}}$-system $X_R^{(\mathbf{cat})}$ (49.1)".

To do this we have to generalize the results of §50 on Kan extensions and homotopical Kan extensions of the identity to systems (§48) and as systems consist not only of categories and functors, but also natural transformations we first have to consider

51.1. A notion of conjugation for Kan extensions. Given two functors $g_1, g_2 \colon Y \to X$, a natural transformation $v \colon g_2 \to g_1$ and terminal (resp. initial) Kan extensions of the identity f_1 and f_2 along g_1 and g_2 respectively we

(i) will refer to the *unique* natural transformation $u \colon f_1 \to f_2$ for which the diagram

$$\begin{array}{ccc} f_1 g_2 & \xrightarrow{v} & f_1 g_1 \\ u \downarrow & & \downarrow e_1 \\ f_2 g_2 & \xrightarrow{e_2} & 1_Y \end{array} \quad (\text{resp.} \quad \begin{array}{ccc} 1_Y & \xrightarrow{e_1} & f_1 g_1 \\ e_2 \downarrow & & \downarrow u \\ f_2 g_2 & \xrightarrow{v} & f_2 g_1 \end{array})$$

commutes as the **conjugate** of v with respect to these Kan extensions

and note that it follows from the diagrams preceding 39.10(v) and 50.1(ii) that

51. HOMOTOPICAL COCOMPLETENESS AND COMPLETENESS (GENERAL CASE) 167

(ii) *if g_1 and g_2 both have left (resp. right) adjoints, this conjugate of v with respect to these Kan extensions is also the conjugate of v with respect o the associated (50.1(ii)) adjunctions.*

Next we also note that the counit and the unit of an adjunction $F \leftrightarrow F'$ between a left $\boldsymbol{C}$-system F and a right $\boldsymbol{C}^{\mathrm{op}}$-system F' (48.8) are compatible with the composers of F and F' in the sense that they are what we will call

51.2. Dinatural transformations. Given a category $\boldsymbol{C}$, a left $\boldsymbol{C}$-system F and a right $\boldsymbol{C}^{\mathrm{op}}$-system F' (48.1) such that $FD = F'D$ for every object $D \in \boldsymbol{C}$, a **dinatural transformation**

$$e \colon FF' \longrightarrow 1 \qquad (\text{resp. } e \colon 1 \longrightarrow F'F)$$

will be a function e which assigns to every map $u \colon A \to B \in \boldsymbol{C}$ a natural transformation

$$eu \colon (Fu)(F'u^{\mathrm{op}}) \longrightarrow 1_{FB} \qquad (\text{resp. } eu^{\mathrm{op}} \colon 1_{FA} \longrightarrow (F'u^{\mathrm{op}})(Fu))$$

which is compatible with the composers of F and F' in the sense that for every composable pair of maps $u \colon A \to B$ and $v \colon B \to D \in \boldsymbol{C}$, the diagram

$$\begin{array}{ccc}
(Fv)(Fu)F'(u^{\mathrm{op}}v^{\mathrm{op}}) & \xrightarrow{F(v,u)} & F(vu)F'(vu)^{\mathrm{op}} \\
{\scriptstyle F'(u^{\mathrm{op}},v^{\mathrm{op}})}\downarrow & & \downarrow{\scriptstyle e(vu)} \\
(Fv)(Fu)(F'u^{\mathrm{op}})(F'v^{\mathrm{op}}) & \xrightarrow{(ev)(eu)} & 1_{FD}
\end{array}$$

$$\left(\text{resp.} \quad \begin{array}{ccc}
1_{FA} & \xrightarrow{(ev^{\mathrm{op}})(eu^{\mathrm{op}})} & (F'u^{\mathrm{op}})(F'v^{\mathrm{op}})(Fv)(Fu) \\
{\scriptstyle e(vu)^{\mathrm{op}}}\downarrow & & \downarrow{\scriptstyle F(v,u)} \\
F'(vu)^{\mathrm{op}}F(vu) & \xrightarrow{F'(u^{\mathrm{op}},v^{\mathrm{op}})} & (F'u^{\mathrm{op}})(F'v^{\mathrm{op}})F(vu)
\end{array}\right)$$

commutes. Using 39.10(iv) and the diagrams preceding 39.10(v) one then can show that

(i) *for every adjunction $F \leftrightarrow F'$ between a left $\boldsymbol{C}$-system F and a right $\boldsymbol{C}^{\mathrm{op}}$-system F' (48.8) its counit and unit are dinatural transformations*

$$FF' \longrightarrow 1 \quad \text{and} \quad 1 \longrightarrow F'F$$

and

(ii) *for every left $\boldsymbol{C}$-system F, right $\boldsymbol{C}^{\mathrm{op}}$-system F' and dinatural transformation*

$$e \colon FF' \longrightarrow 1 \qquad (\text{resp. } e \colon 1 \longrightarrow F'F)$$

there is at most one adjunction $F \leftrightarrow F'$ with e as its counit (resp. unit).

Now we are ready to deal with

51.3. Kan extensions and homotopical Kan extensions along systems.
Given a category C and a right (resp. a left) C^{op}-system G, let

$$(-G \downarrow 1) \qquad (\text{resp. } (1 \downarrow -G))$$

denote the homotopical category which has as objects the pairs (F, e) consisting of a left (resp. a right) C-system F and a dinatural transformation (51.2)

$$e \colon FG \longrightarrow 1 \qquad (\text{resp. } e \colon 1 \longrightarrow FG)$$

and as maps and weak equivalences $(F_1, e_1) \to (F_2, e_2)$ the maps and weak equivalences $t \colon F_1 \to F_2$ (48.2) such that, for every map $u \colon A \to B \in C$, the diagram

$$
\begin{array}{c}
(F_1 u)(G u^{\mathrm{op}}) \\
\downarrow tu \quad \overset{e_1 u}{\searrow} \\
\quad\quad\quad 1 \\
\overset{e_2 u}{\nearrow} \\
(F_2 u)(G u^{\mathrm{op}})
\end{array}
\qquad (\text{resp. }
\begin{array}{c}
(F_1 u)(G u^{\mathrm{op}}) \\
\overset{e_1 u}{\nearrow} \quad\quad \downarrow tu \\
1 \\
\underset{e_2 u}{\searrow} \\
(F_2 u)(G u^{\mathrm{op}})
\end{array}
)
$$

commutes, and let

$$(-G \downarrow 1)_{\mathrm{w}} \subset (-G \downarrow 1) \qquad (\text{resp. } (1 \downarrow -G)_{\mathrm{w}} \subset (1 \downarrow -G))$$

denote the full subcategory spanned by the pairs (F, e) for which F is homotopical (48.1).

We then define
(i) a **terminal** (resp. an **initial**) **Kan extension** along G as an object

$$(F, e) \in (-G \downarrow 1) \qquad (\text{resp. } (F, e) \in (1 \downarrow -G))$$

such that
(i)′ (F, e) is a terminal (resp. an initial) object, and
(i)″ for every map $u \colon A \to B \in C$, the pair (Fu, eu) is a terminal (resp. an initial) Kan extension of the identity along $G^{u^{\mathrm{op}}}$ (47.1)

and note that, in view of 37.2(v)
(ii) *such Kan extensions, if they exist, are categorically unique* (37.1).

Similarly we define
(iii) a **homotopically terminal** (resp. **initial**) **Kan extension** along G as an object

$$(K, d) \in (-G \downarrow 1)_{\mathrm{w}} \qquad (\text{resp. } (k, d) \in (1 \downarrow -G)_{\mathrm{w}})$$

such that
(iii)′ (K, d) is a homotopically terminal (resp. initial) object, and
(iii)″ for every map $u \colon A \to B \in C$, the pair (Ku, du) is a homotopically terminal (resp. initial) Kan extension of the identity along Gu^{op} (50.2)

and note that it follows from 37.6(vi), 38.3(i) and 50.2(i) that
(iv) *if one object of $(-G \downarrow 1)_{\mathrm{w}}$ (resp. $(1 \downarrow -G)_{\mathrm{w}}$) satisfies (iii)′ and (iii)″, then every object that satisfies one of (iii)′ and (iii)″ also satisfies the other*

so that (37.6(vi))
(v) *such homotopical Kan extensions, if they exist, are homotopically unique* (37.5).

One then has the following

51.4. Sufficient conditions for existence of Kan extensions and homotopical Kan extensions along systems. It follows from 50.1(ii) and 51.1(ii) that

(i) *for every right (resp. left) $\boldsymbol{C}^{\mathrm{op}}$-system G which has left (resp. right) adjoints (48.8), a pair (F,e) is a terminal (resp. an initial) Kan extension along G iff there exists an adjunction $F \leftrightarrow G$ (resp. $G \leftrightarrow F$) with e as its counit (resp. unit),*

so that

(i)′ *a sufficient condition for the existence of a terminal (resp. an initial) Kan extension along G is the existence of a left (resp. a right) adjoint of G.*

Similarly it follows from 50.2(ii) that

(ii) *for every right (resp. left) $\boldsymbol{C}^{\mathrm{op}}$-system G, every terminal (resp. initial) Kan extension (F,e) along G gives rise to a 1-1 correspondence between the left (resp. the right) approximations of F (48.5) and the homotopically terminal (resp. initial) Kan extensions along G, which is induced by the isomorphism (48.2)*

$$(\boldsymbol{C}_{\mathrm{L}}\text{-}\mathbf{syst}_{\mathrm{w}} \downarrow F) \approx (-G \downarrow 1)_{\mathrm{w}} \qquad (\textit{resp. } (F \downarrow \boldsymbol{C}_{\mathrm{R}}\text{-}\mathbf{syst}_{\mathrm{w}}) \approx (1 \downarrow -G)_{\mathrm{w}})$$

which sends an object (K,a) to the pair (K,d) in which d denotes the dinatural transformation (51.2)

$$(Ku)(Gu^{\mathrm{op}}) \xrightarrow{au} (Fu)(Gu^{\mathrm{op}}) \xrightarrow{eu} 1_{GB}$$

$$(\textit{resp. } 1_{GB} \xrightarrow{eu} (Fu)(Gu^{\mathrm{op}}) \xrightarrow{au} (Ku)(Gu^{\mathrm{op}}))$$

which together with (i)′ and 48.7 implies that

(iii) *a sufficient condition for the existence of a homotopically terminal (resp. initial) Kan extension along G is that there exists a terminal (resp. initial) Kan extension along G which is left (resp. right) deformable (48.6).*

and suggests that, as in 50.2(iv)

(iv) *given a a right (resp. a left) $\boldsymbol{C}^{\mathrm{op}}$-system G, a terminal (resp. an initial) Kan extension (F,e) along G and a pair consisting of a left (resp. a right) approximation (K,a) of F and the corresponding (see (v) above) homotopically terminal (resp. initial) Kan extension (K,d) along G, say that (K,a) is the F-**presentation** of (K,d).*

Finally we are ready to extend to not necessarily cocomplete or complete homotopical categories the results of §49 on

51.5. Homotopical cocompleteness and completeness. Given a homotopical category $\boldsymbol{X}$, we will say that $\boldsymbol{X}$ is **homotopically cocomplete** (resp. **complete**) if there exists a **homotopical colimit** (resp. **limit**) **system** on $\boldsymbol{X}$, which we define as a *homotopically terminal* (resp. *initial*) *Kan extension* (51.3(iii)) along the right (resp. the left) $\mathbf{cat}^{\mathrm{op}}$-system $\boldsymbol{X}_R^{(\mathbf{cat})}$ (resp. $\boldsymbol{X}_L^{(\mathbf{cat})}$) (49.1). Then (51.3(v))

(i) *such homotopical colimit (resp. limit) systems on $\boldsymbol{X}$, if they exist, are homotopically unique (37.5)* while 49.1 and 51.4(ii)′ imply that

(ii) *a sufficient condition for the homotopical cocompleteness (resp. completeness) of X, i.e. the existence of homotopical colimit (resp. limit) systems on X,*

is that

(ii)′ *there exists a terminal (resp. initial) Kan extension along $X_R^{(\mathbf{cat})}$ (resp. $X_L^{(\mathbf{cat})}$) which is left (resp. right) deformable (48.6)*

Moreover it follows from 49.1 and 51.4(i) and (iii) that, *if X is cocomplete (resp. complete)*, then

(iii) *there exist homotopical colimit (resp. limit) systems on X*

iff

(iii)′ *there exist homotopy colimit (resp. limit) systems on X (49.2)*

and

(iii)″ *if (iii) and (iii)′ hold then the homotopy colimit (resp. limit) systems on X are exactly the $\operatorname{colim}^{(\mathbf{cat})}$- (resp. $\operatorname{lim}^{(\mathbf{cat})}$-) presentations (51.4(iv)) of the homotopical colimit (resp. limit) systems on X*

so that

(iv) *for cocomplete (resp. complete) homotopical categories the above notion of homotopical cocompleteness (resp. completeness) coincides with the one we considered in §49.*

Index

Page numbers in bold point to the definitions of the indexed terms.

adjunctions, 122–123
  compatibility with (co)limit functors, 57, 151
  counits of, **122**, 122–123, 135–138
  deformable, 14–15, 37, **51**, 51–54, 119–122, **133**
    derived adjunctions of, 8, 14, 36–37, **46**, **53**, 44–54, 121, **139**, 138–143, *see also* derived adjunctions
  partial, **44**, **53**, **133**, 133–138
  Quillen, **40**, *see also* Quillen adjunctions
  units of, **122**, 122–123, 135–138
adjunctions of systems, **157**
  and dinatural transformations, 167
  and Kan extensions along systems, 169
  categorical uniqueness, 157
  counits of, **157**
  deformable, 157
    derived adjunctions of, **158**
  locally deformable, 157
  sufficient conditions for existence, 157
  units of, **157**
all or none proposition for compositions, 131
alternate description of homotopy categories, 100
approximations, 13, 35–37, 42–44, **51**, 84, 120–122, **127**, 126–130
  and total derived functors, 128
  compositions of, 44, 120–121, 130–133
  homotopical uniqueness of, 42, 51, 127
  of deformable functors, 51, 127
  of homotopical functors, 128
  of Kan extensions
    and homotopical Kan extensions, 163
  of natural transformations, **129**, 129–130
  of Quillen functors, **42**
  sufficient conditions for composability, 52
  sufficient conditions for existence, 51, 127
approximations of systems, **154**
  homotopical uniqueness of, 154
  sufficient conditions for existence, 155
arrow categories, **101**, 101–112
  T-diagrams of, 102

axiom
  factorization, 4, 19, **26**
  lifting, **26**
  limit, 3, 19, **25**
  retract, 19–20, **26**
  two out of six, **26**
  two out of three, 19–20

canonical natural isomorphisms, **45**
canonical natural transformations, **139**, 139–142
canonically isomorphic objects, 10, **38**, 82, 92, **112**, *see also* categorical uniqueness
canonically weakly equivalent objects, 10, **39**, 83, 92, **114**, *see also* homotopical uniqueness
CAT, **95**, 97–99
cat, **95**
CAT$_w$, **97**, 97–99
cat-systems, *see also* systems
  left, **62**, 62–65, **73**
  right, **62**, 62–65, **73**
categorical uniqueness, **38**, 38–39, 92–93, **112**, 112–114
  of adjoints of systems, 157
  of colimit systems, 160
  of initial objects, 38, 92, 113
  of limit systems, 160
  of terminal objects, 38, 92, 113
categorically
  contractible categories, 10, **38**, 82, 92, **113**
  full subcategories, **38**, 82, 92, **112**
  unique objects, 10, **38**, 82, 92, **112**, *see also* categorical uniqueness
categories, 4, **23**, 22–23, 78–79, 89–90, **95**, 95
  arrow, **101**, 101–112
  categorically contractible, 10, **38**, 82, 92, **113**
  closed model, 3–4, 19, 27–28
  cocomplete, **25**, **58**, 58, **159**, *see also* cocompleteness
  complete, **25**, **58**, 58, **159**, *see also* completeness

171

connected components of, **102**
diagram, 22, **24**, **97**
functor, 22, **24**, **97**
homotopical, 11–12, **23**, 20–25, **77**, 79–80, 90–91, **96**, 96–101
homotopically cocomplete, 16–17, **73**, **160**, **169**, *see also* homotopical cocompleteness
homotopically complete, 16–17, **73**, **160**, **169**, *see also* homotopical completeness
homotopically contractible, 10, 36, **39**, 83, 92, **114**
homotopy, 4–8, 21–22, *see also* homotopy categories
indexing, **97**
locally small, 4, **23**, 79, 89–90, **95**
model, 3–4, 19–20, **25**, 25–29, *see also* model categories
n-arrow, **101**, *see also* arrow categories
of simplices, **67**, 67–72
of weak equivalences, **96**
Reedy, **65**
Reedy model, 65–72
simplicial, 105
small, **23**, 79, 89–90, **95**
small $\mathcal{U}$-, **22**, 90, **94**, 94–95
$\mathcal{U}$-, **22**, 90, **94**, **94**, 94–95
underlying, **96**
we-, 4–5
with cofibrant constants, **67**, 67–68
with fibrant constants, **67**, 67–68
category of types, **101**
classical homotopy categories, **31**
classifying space, **104**
closed model categories, 3–4, 19, 27–28
closure properties of model categories, 28
cocomplete categories, **25**, **58**, 58, **159**, *see also* cocompleteness
homotopically, 16–17, **73**, **160**, **169**, *see also* homotopical cocompleteness
cocompleteness, **58**, 58, 159–160
and colimit systems, 58, 160
homotopical, 16–17, 56, 62–65, **73**, 86–87, 159–161, **169**, *see also* homotopical cocompleteness
cofibrant constants, **67**, 67–68
cofibrant fibrant objects, **30**
cofibrant objects, **30**
cofibrations, **25**
characterization of, 28
Reedy, **65**
trivial, **26**
characterization of, 28
colim$^{(\mathbf{cat})}$, 58
colim^D, 57, **148**
colim^u, 57, **149**
colim$^{(v,u)}$, 57
colimit description

of Grothendieck constructions, 105
of homotopy categories, 81, 102
colimit functors, **57**, 57–58, 85, **148**
compatibility with left adjoints, 57, 151
deformability result for, 56, 61
homotopical, 15–17, 85–87, 165–166, *see also* homotopical colimit functors
homotopy, 8–11, 15–17, 55–56, **59**, 59–62, **72**, 85–87, 148–152, *see also* homotopy colimit functors
colimit systems, **58**, 86, **160**
and cocompleteness, 58, 160
categorical uniqueness of, 160
homotopical, 87, **169**, *see also* homotopical colimit systems
homotopy, **63**, 63–65, **73**, 86, **160**, *see also* homotopy colimit systems
compatibility
of adjoints with (co)limit functors, 57, 151
complete categories, **25**, **58**, 58, **159**, *see also* completeness
homotopically, 16–17, **73**, **160**, **169**, *see also* homotopical completeness
completeness, **58**, 58, 159–160
and limit systems, 58, 160
homotopical, 16–17, 62–65, **73**, 159–161, **169**, *see also* homotopical completeness
components
connected, **102**
composers, **152**
of **cat**-systems, **62**
composition functors, **43**, **130**, **164**, *see also* compositions
compositions
all or none proposition, 131
of approximations, 120–121, 130–133
of approximations of Quillen functors, 44
of deformable functors, 52–53, 84–85, 131–133
of derived adjunctions, 46, 142
of homotopical colimit functors, 165
of homotopical Kan extensions, 164
of homotopical limit functors, 165
of homotopy colimit functors, 60, 72, 150
of homotopy limit functors, 60, 72, 150
of Kan extensions, 164
of partial adjunctions, **44**, 134
conjugate pairs of natural transformations, **123**, 123, **166**
deformable, **133**, 141
connected components, **102**
constant diagram functors, **57**, **148**
contractible categories
categorically, 10, **38**, 82, 92, **113**
homotopically, 10, 36, **39**, 83, 92, **114**
counits
of adjunctions, **122**, 122–123, 135–138
of adjunctions of systems, **157**

and dinatural transformations, 167
of Kan extensions, **118**, **162**

D-colimit functors, **57**, 85, **148**, *see also* colimit functors
D-limit functors, **57**, **148**, *see also* limit functors
deformability result
  for colimit and limit functors, 56, 61
  for Quillen functors, 41
deformable adjunctions, 14–15, 37, **51**, 51–54, 119–122, **133**
  derived adjunctions of, 8, 14, 36–37, **46**, **53**, 44–54, 121, **139**, 138–143, *see also* derived adjunctions
  Quillen condition for, 38, **54**, 121, **143**, 143–145
deformable adjunctions of systems, 157
  derived adjunctions of, **158**
deformable conjugate pairs of natural transformations, **133**, 141
deformable functors, 7, 13–15, **51**, 51–54, 83–85, **124**, 119–126
  approximations of, 51, 127
  compositions of, 52–53, 84–85, 131–133
deformable systems, 87, **155**, *see also* systems
deformation retracts, 7, 24, 83, 119 **124**
  and model categories, 30
deformations, 7, **24**, 83, 119, **124**, *see also* f-deformations
degree function, **65**
derived adjunctions, 8, 14, 36–37, **46**, **53**, 44–54, 121, **139**, 138–143
  compositions of, 46, 142
  conjugations between, 141
  of deformable adjunctions of systems, **158**
  of homotopy colimit functors, 59, 149
  of homotopy colimit systems, 161
  of homotopy limit functors, 59, 149
  of homotopy limit systems, 161
derived functors
  total left, 5, 7, **128**
  total right, 5, 7, **128**
diagram categories, 22, **24**, **97**
  homotopical, **97**
diagrams, **96**
  restricted, **69**, 69–72
  virtually cofibrant, **69**, 69–72
  virtually fibrant, **69**, 69–72
dinatural transformations, **167**, 167–168
  and adjunctions of systems, 167
  and counits of adjunctions, 167
  and units of adjunctions, 167

embedding **CAT** in **CAT**$_w$, **98**
enrichment
  Grothendieck, **104**

f-deformation retracts, **124**
  maximal, 125
F-deformations, **155**
f-deformations, 7, **124**
  homotopical uniqueness of, 125
F-presentations, **169**
  of homotopical Kan extensions along systems, **169**
f-presentations, **163**
  of homotopical Kan extensions, **163**
factorization axiom, 4, 19, **26**
fibrant constants, **67**, 67–68
fibrant objects, **30**
fibrations, **25**
  characterization of, 28
  Reedy, **65**
  trivial, **26**
    characterization of, 28
full subcategories
  categorically, **38**, 82, 92, **112**
  homotopically, **39**, 83, 92, **114**
Fun, **24**, **97**
Fun$_w$, **24**, **97**
functor categories, 22, **24**, **97**
  homotopical, **97**
functors
  colimit, **57**, 57–58, 85, **148**, *see also* colimit functors
  composition, **43**, **130**, **164**, *see also* compositions
  constant diagram, **57**, **148**
  D-colimit, **57**, 85, **148**, *see also* colimit functors
  D-limit, **57**, **148**, *see also* limit functors
  deformable, 7, 13–15, **51**, 51–54, 83–85, **124**, 119–126, *see also* deformable functors
  deformable pairs of, 37, **52**, **131**, *see also* pairs of functors
  forgetful, **97**
  homotopical, 12, **24**, 24–25, 79–80, 90, **96**
  homotopical categories of, **97**
  homotopical colimit, 15–17, 85–87, 165–166, *see also* homotopical colimit functors
  homotopical limit, 15–17, 165–166, *see also* homotopical limit functors
  homotopical u-colimit, 87, **165**, *see also* homotopical colimit functors
  homotopical u-limit, **165**, *see also* homotopical limit functors
  homotopy colimit, 8–11, 15–17, 55–56, **59**, 59–62, **72**, 85–87, 148–152, *see also* homotopy colimit functors
  homotopy D-colimit, 8–11, **59**, 85, **148**, *see also* homotopy colimit functors
  homotopy D-limit, **59**, **148**, *see also* homotopy limit functors

homotopy limit, 15–17, **59**, 59–62, **72**, 148–152, *see also* homotopy limit functors
homotopy u-colimit, **59**, **72**, 85, **149**, *see also* homotopy colimit functors
homotopy u-limit, **59**, **72**, **149**, *see also* homotopy limit functors
induced diagram, **57**, **148**
initial projection, **68**, 68–72
left deformable, 7, 13–15, 37, **51**, 51–54, **124**, 119–126, *see also* deformable functors
left deformable pairs of, 37, **52**, **131**, *see also* pairs of functors
left Quillen, 35, **40**, 40–54, *see also* Quillen functors
limit, **57**, 57–58, *see also* limit functors
localization, 4, 22, **24**, **29**
locally deformable pairs of, 37, **52**, **131**, *see also* pairs of functors
locally left deformable pairs of, 37, **52**, **131**, *see also* pairs of functors
locally right deformable pairs of, 37, **52**, **131**, *see also* pairs of functors
naturally weakly equivalent, **24**, 90, **96**
projection, **68**, 68–72
Quillen, 35, **40**, 40–54, *see also* Quillen functors
right deformable, 7, 13–15, 37, **51**, 51–54, **124**, 119–126, *see also* deformable functors
right deformable pairs of, 37, **52**, *see also* pairs of functors
right Quillen, 35, **40**, 40–54, *see also* Quillen functors
terminal projection, **68**, 68–72
total left derived, 5, 7, **128**
total right derived, 5, 7, **128**
u-colimit, **57**, 85, **148**, *see also* colimit functors
u-limit, **57**, **148**, *see also* limit functors

γ, **24**, **99**
Gr, **103**, 103–107
Grothendieck construction, **103**, 103–107
  and simplicial localizations, 105
  colimit description of, 105
Grothendieck description of homotopy categories, 104
Grothendieck enrichment, 91, **104**

h-deformation retracts, **126**
h-deformations, **126**
hammock localizations, *see also* simplicial localizations
higher universes, **94**
Ho, **24**, **98**, 98–104
homotopic maps, **31**, 45
homotopical **cat**-systems, **63**

homotopical categories, 11–12, **23**, 20–25, **77**, 79–80, 90–91, **96**, 96–101
  homotopical equivalences of, **24**, 80, 91, **96**
  homotopy categories of, 12–13, **24**, 24–25, 80–82, 91–92, **98**, 98–107, *see also* homotopy categories
  locally small, **96**
  maximal, 80, **97**
  minimal, 80, **98**
  of functors, **97**
  of homotopical functors, **97**
  saturated, **25**, 37, 38, 52, 54, 73, 82, 87, 92, **99**, 121, 122, 132, 144, 150, *see also* saturation
  small, **96**
  weak equivalences in, 11, **23**, 90, **96**
homotopical cocompleteness, 16–17, 56, 62–65, **73**, 86–87, 159–161, **169**, *see also* homotopical colimit systems, homotopy colimit system
  of model categories, 64
  sufficient conditions for, 73, 161, 170
homotopical colimit functors, 15–17, 85–87, 165–166
  and homotopy colimit functors, 165
  homotopical uniqueness of, 165
  sufficient conditions for composability, 165
  sufficient conditions for existence, 165
homotopical colimit systems, 87, **169**
  and homotopy colimit systems, 170
  homotopical uniqueness of, 169
  sufficient conditions for existence, 170
homotopical compatibility
  of deformable adjoints with homotopy (co)limit functors, 72, 151–152
  of Quillen functors with homotopy (co)limit functors, **56**, **60**
homotopical completeness, 16–17, 62–65, **73**, 159–161, **169**, *see also* homotopical limit systems, homotopy limit system
  of model categories, 64
  sufficient conditions for, 73, 161, 170
homotopical diagram categories, **97**
  3-arrow calculi on, 108
  and saturation, 25, 99
homotopical equivalences of homotopical categories, **24**, 80, 91, **96**
homotopical functor categories, **97**
  3-arrow calculi on, 108
  and saturation, 25, 99
homotopical functors, 12, **24**, 24–25, 79–80, 90, **96**
  approximations of, 128, *see also* approximations
  homotopical categories of, **97**
homotopical inverses, **24**, **96**

homotopical Kan extensions, 87, 93, **118**, **163**
  and approximations of Kan extensions, 163
  homotopical uniqueness of, 118, 163
  presentations of, **163**
  sufficient conditions for composability, 164
  sufficient conditions for existence, 163
homotopical Kan extensions along systems, **168**
  and approximations of Kan extensions along systems, 169
  homotopical uniqueness of, 168
  presentations of, **169**
  sufficient conditions for existence, 169
homotopical limit functors, 15–17, 165–166
  and homotopy limit functors, 165
  homotopical uniqueness of, 165
  sufficient conditions for composability, 165
  sufficient conditions for existence, 165
homotopical limit systems, **169**
  and homotopy limit systems, 170
  homotopical uniqueness of, 169
  sufficient conditions for existence, 170
homotopical structures, **23**, **96**
homotopical subcategories, **96**
homotopical systems, **153**
  homotopy systems of, **154**
homotopical u-colimit functors, 87, **165**, *see also* homotopical colimit functors
homotopical u-limit functors, **165**, *see also* homotopical limit functors
homotopical uniqueness, 9–10, **39**, 39–40, 82–83, 92–93, **114**, 112–118
  of approximations, 42, 51, 127, 154
  of f-deformations, 125
  of homotopical colimit functors, 165
  of homotopical colimit systems, 169
  of homotopical Kan extensions, 118, 163
  of homotopical Kan extensions along systems, 168
  of homotopical limit functors, 165
  of homotopical limit systems, 169
  of homotopically initial objects, 40, 93, 115
  of homotopically terminal objects, 40, 93, 115
  of homotopy colimit functors, 59, 72, 149
  of homotopy colimit systems, 64, 160
  of homotopy limit functors, 59, 72, 149
  of homotopy limit systems, 64, 160
homotopical version, 80, **98**
homotopically
  cocomplete categories, 16–17, **73**, **160**, **169**, *see also* homotopical cocompleteness
  complete categories, 16–17, **73**, **160**, **169**, *see also* homotopical completeness
  contractible categories, 10, 36, **39**, 83, 92, **114**

full subcategories, **39**, 83, 92, **114**
homotopically initial Kan extensions, **118**, **163**, *see also* homotopical Kan extensions
homotopically initial Kan extensions along left systems, **168**, *see also* homotopical Kan extensions along systems
homotopically initial objects, 13, **39**, 39–40, 83, 93, **116**, 115–118
  homotopical uniqueness of, 40, 93, 115
  motivation, 115
homotopically terminal Kan extensions, **118**, **163**, *see also* homotopical Kan extensions
homotopically terminal Kan extensions along right systems, **168**, *see also* homotopical Kan extensions along systems
homotopically terminal objects, 9, 13, **39**, 39–40, 83, 93, **116**, 115–118
  homotopical uniqueness of, 40, 93, 115
  motivation, 115
homotopically unique objects, 10, 35, **39**, 83, 92, **114**, *see also* homotopical uniqueness
homotopically universal properties, **40**, **115**
homotopy categories, 4–8, 21–22
  alternate description of, 100
  classical, **31**
  colimit description of, 81, 102
  descriptions of, 98–112
  Grothendieck description of, 104
  of homotopical categories, 12–13, **24**, 24–25, 80–82, 91–92, **98**, 98–107
  of model categories, **29**, 29–32
  3-arrow description of, 5, 21, 32, 33, 81, 91, 109
homotopy colimit functors, 8–11, 15–17, 55–56, **59**, 59–62, **72**, 85–87, 148–152
  and homotopical colimit functors, 165
  compositions of, 60, 72, 150
  derived adjunctions of, 59, 149
  existence on model categories, 59
  homotopical compatibility with left deformable left adjoints, 72, 151–152
  homotopical compatibility with left Quillen functors, **56**, **60**
  homotopical uniqueness of, 59, 72, 149
  sufficient conditions for composability, 72, 150
  sufficient conditions for existence, 72, 149
homotopy colimit systems, **63**, 63–65, **73**, 86, **160**
  and homotopical colimit systems, 170
  derived adjunctions of, 161
  homotopical uniqueness of, 64, 160
  sufficient conditions for existence, 73, 161
homotopy D-colimit functors, 8–11, **59**, 85, **148**, *see also* homotopy colimit functors

homotopy D-limit functors, **59**, **148**, *see also* homotopy limit functors
homotopy equivalences, **31**
homotopy inverses, **31**
homotopy limit functors, 15–17, **59**, 59–62, **72**, 148–152
  and homotopical limit functors, 165
  compositions of, 60, 72, 150
  derived adjunctions of, 59, 149
  existence on model categories, 59
  homotopical compatibility with right deformable right adjoints, 72, 151–152
  homotopical compatibility with right Quillen functors, 60
  homotopical uniqueness of, 59, 72, 149
  sufficient conditions for composability, 72, 150
  sufficient conditions for existence, 72, 149
homotopy limit systems, **63**, 63–65, **73**, **160**
  and homotopical limit systems, 170
  derived adjunctions of, 161
  homotopical uniqueness of, 64, 160
  sufficient conditions for existence, 73, 161
homotopy relations, 30–32
homotopy systems of homotopical systems, **154**
homotopy u-colimit functors, **59**, **72**, 85, **149**, *see also* homotopy colimit functors
homotopy u-limit functors, **59**, **72**, **149**, *see also* homotopy limit functors

indexing categories, **97**
induced diagram functors, **57**, **148**
initial Kan extensions, **118**, 128, **162**, *see also* Kan extensions
initial Kan extensions along left systems, **168**, *see also* Kan extensions along systems
initial objects
  categorical uniqueness of, 38, 92, 113
  homotopically, 13, **39**, 39–40, 83, **116**, 115–118, *see also* homotopically initial objects
initial projection functors, **68**, 68–72
inverses
  homotopical, **24**, **96**
  homotopy, **31**
invertibility property
  weak, **23**, **96**

Kan extensions, 87, **118**, 128, **162**
  counits of, **118**, **162**
  homotopical, 87, 93, **118**, **163**, *see also* homotopical Kan extensions
  sufficient conditions for composability, 164
  sufficient conditions for existence, 162
  units of, **118**, **162**
Kan extensions along systems, **168**
  and adjunctions of systems, 169
  homotopical, *see also* homotopical Kan extensions along systems
  sufficient conditions for existence, 169
Ken Brown's lemma, 41

latching objects, **66**
left adjoints
  compatibility with colimit functors, 57, 151
  of left systems, **157**, *see also* adjunctions of systems
left approximations, 13, 35–37, 42–44, **51**, 84, 120–122, **127**, 126–130, *see also* approximations
left **cat**-systems, **62**, 62–65, **73**, *see also* systems
left deformable functors, 7, 13–15, 37, **51**, 51–54, 84, **124**, 119–126, *see also* deformable functors
left deformable left adjoints
  homotopical compatibility with homotopy colimit functors, 151–152
left deformable natural transformations, **126**
left deformable pairs of functors, 37, **52**, 84, 120, **131**, *see also* pairs of functors
left deformable systems, 87, **155**, *see also* systems
left deformation retracts, 7, 24, 83, 119, **124**
left deformations, 7, **24**, 83, 119, **124**
left f-deformation retracts, **124**
  maximal, 125
left F-deformations, **155**
left f-deformations, 7, **124**
  homotopical uniqueness of, 125
left h-deformation retracts, **126**
left h-deformations, **126**
left homotopic maps, **30**, 45
left lifting property, **26**
left Quillen equivalences, 36–37, **49**, 48–50, *see also* Quillen equivalences
left Quillen functors, 35, **40**, 40–54, *see also* Quillen functors
left systems, **152**, *see also* systems
length
  of a zigzag, **98**
lifting axiom, **26**
$\lim^{(\mathbf{cat})}$, 58
$\lim^{D}$, 57, **148**
$\lim^{u}$, 57, **149**
$\lim^{(v,u)}$, 57
limit axiom, 3, 19, **25**
limit functors, **57**, 57–58, **148**
  compatibility with right adjoints, 57, 151
  deformability result for, 56, 61
  homotopical, 15–17, 165–166, *see also* homotopical limit functors
  homotopy, 15–17, **59**, 59–62, **72**, 148–152, *see also* homotopy limit functors
limit systems, **58**, **160**

and completeness, 58, 160
categorical uniqueness of, 160
homotopical, **169**, *see also* homotopical limit systems
homotopy, **63**, 63–65, **73**, **160**, *see also* homotopy limit systems
local left F-deformations, **155**
local right F-deformations, **155**
localization, 4, 22
  simplicial, 105
localization functors, 4, 22, **24**, **29**, **99**
locally deformable adjunctions of systems, 157, *see also* adjunctions of systems
locally left deformable pairs of functors, 37, **52**, 84, 120, **131**, *see also* pairs of functors
locally left deformable systems, 87, **155**, *see also* systems
locally right deformable pairs of functors, 37, **52**, 84, 120, **131**, *see also* pairs of functors
locally right deformable systems, **155**, *see also* systems
locally small
  categories, 4, **23**, 79, 89–90, **95**
  homotopical categories, **96**

maps
  between **cat**-systems, **63**
  between left systems, **153**
  between right systems, **153**
  homotopic, **31**, 45
  left homotopic, **30**, 45
  right homotopic, **31**, 45
matching objects, **66**
maximal
  f-deformation retracts, 125
  homotopical categories, 80, **97**
  model structures, **29**
  structure functors, **97**
minimal
  homotopical categories, 80, **98**
  model structures, **29**
  structure functor, **98**
model categories, 3–4, 19–20, **25**, 25–29
  and deformation retracts, 30
  closed, 3–4, 19, 27–28
  closure properties, 28
  colimit systems on, 64
  deformability result for colimit and limit functors, 56, 61
  homotopical cocompleteness of, 64
  homotopical completeness of, 64
  homotopy categories of, **29**, 29–32
  homotopy colimit functors on, 59
  homotopy limit functors on, 59
  Ken Brown's lemma, 41
  limit systems on, 64

  Reedy, 65–72
  saturation of, 21, 31
  3-arrow calculi of, 6, 34
  weak equivalences in, 20, **25**
model structures, **25**
  maximal, **29**
  minimal, **29**
  Reedy, **65**, 65–68

n-arrow categories, **101**, *see also* arrow categories
natural transformations
  approximations of, **129**, 129–130
  canonical, **139**, 139–142
  conjugate pairs of, **123**, 123, **166**
  deformable, **133**
  deformable, **126**
  di-, **167**, 167–168
natural weak equivalences, **24**, 90, **96**
naturally weakly equivalent functors, **24**, 90, **96**
nerve, **104**

objects
  canonically isomorphic, 10, **38**, 82, 92, **112**, *see also* categorical uniqueness
  canonically weakly equivalent, 10, **39**, 83, 92, **114**, *see also* homotopical uniqueness
  categorically unique, 10, **38**, 82, 92, **112**, *see also* categorical uniqueness
  cofibrant, **30**
  cofibrant fibrant, **30**
  fibrant, **30**
  homotopically initial, 13, **39**, 39–40, 83, 93, **116**, 115–118, *see also* homotopically initial objects
  homotopically terminal, 9, 13, **39**, 39–40, 83, 93, **116**, 115–118, *see also* homotopically terminal objects
  homotopically unique, 10, 35, **39**, 83, 92, **114**, *see also* homotopical uniqueness
  latching, **66**
  matching, **66**
  weakly equivalent, **23**, **96**

pairs of functors
  deformable, 37, **52**, 84, 120, **131**
  locally deformable, 37, **52**, 84, 120, **131**
  sufficient conditions for deformability, 52, 132
partial adjunction functors, **133**
partial adjunction isomorphisms, **44**, **53**, **134**, 140–143
partial adjunctions, **44**, **53**, **133**, 133–138
  compositions of, **44**, 134
  naturality of, 134
presentations
  F-, **169**

f-, **163**
   of homotopical Kan extensions, **163**
   of homotopical Kan extensions along systems, **169**
projection functors, **68**, 68–72
property
   homotopically universal, **40**, **115**
   left lifting, **26**
   right lifting, **26**
   two out of six, 10–11, 19, **23**, 79, 90, **96**, 110, 117, 125
   two out of three, 4, 11, **23**, 79, 90, **96**
   universal, **38**, **113**
   weak invertibility, **23**, 79, **96**

Quillen adjunctions, **40**
   and Reedy model structures, 66
   derived adjunctions of, 46
   Quillen condition for, **49**
   Quillen conditions for, **37**
Quillen condition
   for deformable adjunctions, 38, **54**, 121, **143**, 143–145
   for Quillen adjunctions, **37**, **49**
Quillen equivalences, 36–37, **49**, 48–50
   Quillen condition for, **37**, **49**
Quillen functors, 35, **40**, 40–54
   approximations of, **42**
   compositions of approximations, 44
   deformability result for, 41
   existence of approximations, 42
   homotopical compatibility with homotopy (co)limit functors, **56**, **60**

Reedy categories, **65**
Reedy cofibrations, **65**
Reedy fibrations, **65**
Reedy model categories, 65–72
Reedy model structures, **65**, 65–68
   and Quillen adjunctions, 66
   explicit description of, 66
   implicit description of, 66
Reedy weak equivalences, **65**
restricted diagrams, **69**, 69–72
restricted zigzags, 81, **98**, **101**
retract axiom, 19–20, **26**
right adjoints
   compatibility with limit functors, 57, 151
   of right systems, **157**, see also adjunctions of systems
right approximations, 13, 35–37, 42–44, **51**, 84, 120–122, **127**, 126–130, see also approximations
right **cat**-systems, **62**, 62–65, **73**, see also systems
right deformable functors, 7, 13–15, 37, **51**, 51–54, 84, **124**, 119–126, see also deformable functors
right deformable natural transformations, **126**

right deformable pairs of functors, 37, **52**, 84, 120, **131**, see also pairs of functors
right deformable right adjoints
   homotopical compatibility with homotopy limit functors, 151–152
right deformable systems, **155**, see also systems
right deformation retracts, 7, 24, 83, 119, **124**
right deformations, 7, **24**, 83, 119, **124**
right f-deformation retracts, **124**
   maximal, 125
right F-deformations, **155**
right f-deformations, 7, **124**
   homotopical uniqueness of, 125
right h-deformation retracts, **126**
right h-deformations, **126**
right homotopic maps, **31**, 45
right lifting property, **26**
right Quillen equivalences, 36–37, **49**, 48–50, see also Quillen equivalences
right Quillen functors, 35, **40**, 40–54, see also Quillen functors
right systems, **152**, see also systems

saturated homotopical categories, **25**, 37, 38, 52, 54, 73, 82, 87, 92, **99**, 121, 122, 132, 144, 150, see also saturation
saturated systems, **153**
saturation, 5, **25**, 50, **99**, see also saturated homotopical categories
   and homotopical diagram categories, 25, 99
   and homotopical functor categories, 25, 99
   and 3-arrow calculi, 11, 34, 82, 92, 110
   of model categories, 21, 31
sets, **23**, 79, **95**
   simplicial, **104**
   small, **23**, 79, **95**
   $\mathcal{U}$-, **22**, 89, **94**
simplices, **104**
   categories of, **67**, 67–72
simplicial categories, 105
simplicial localizations, 105
   and Grothendieck construction, 105
simplicial sets, **104**
small categories, **23**, 79, 89–90, **95**
small homotopical categories, **96**
small sets, **23**, 79, **95**
small $\mathcal{U}$-categories, **22**, 90, **94**, 94–95
structures
   homotopical, **23**, **96**
   model, **25**
   Reedy model, **65**, 65–68
subcategories
   categorically full, **38**, 82, 92, **112**
   homotopical, **96**
   homotopically full, **39**, 83, 92, **114**

successor universes, **23**, 90, **94**
sufficient conditions for
    homotopical cocompleteness, 73, 161, 170
    homotopical compatibility of deformable adjoints with homotopy (co)limit functors, 72, 151–152
    homotopical completeness, 73, 161, 170
sufficient conditions for composability of
    approximations, 52
    derived adjunctions, 142
        of homotopy colimit functors, 60, 72, 150
        of homotopy limit functors, 60, 72, 150
    homotopical colimit functors, 165
    homotopical Kan extensions, 164
    homotopical limit functors, 165
    homotopy colimit functors, 72, 150
    homotopy limit functors, 72, 150
    Kan extensions, 164
    partial adjunctions, **44**, 134
sufficient conditions for deformability of
    pairs of functors, 52
    systems, 157
sufficient conditions for existence of
    adjoints of systems, 157
    approximations, 51, 127
    approximations of systems, 155
    derived adjunctions of homotopy colimit functors, 149
    derived adjunctions of homotopy limit functors, 149
    homotopical colimit functors, 165
    homotopical colimit systems, 170
    homotopical Kan extensions, 163
    homotopical Kan extensions along systems, 169
    homotopical limit functors, 165
    homotopical limit systems, 170
    homotopy colimit functors, 72, 149
    homotopy colimit systems, 73, 161
    homotopy limit functors, 72, 149
    homotopy limit systems, 73, 161
    Kan extensions, 162
    Kan extensions along systems, 169
systems, **152**, 157
    adjunctions of, **157**, *see also* adjunctions of systems
    approximations of, **154**, *see also* approximations of systems
    colimit, **58**, 86, **160**, *see also* colimit systems
    deformable, 87, **155**
    homotopical, **153**
        homotopy systems of, **154**
    homotopical colimit, 87, **169**, *see also* homotopical colimit systems
    homotopical Kan extensions along, **168**, *see also* homotopical Kan extensions along systems
    homotopical limit, **169**, *see also* homotopical limit systems
    homotopy colimit, **63**, 63–65, **73**, 86, **160**, *see also* homotopy colimit systems
    homotopy limit, **63**, 63–65, **73**, **160**, *see also* homotopy limit systems
    Kan extensions along, **168**, *see also* Kan extensions along systems
    left, **152**
    left **cat**-, **62**, 62–65, **73**
    left deformable, 87, **155**
    limit, **58**, **160**, *see also* limit systems
    locally deformable, 87, **155**
    locally left deformable, 87, **155**
    locally right deformable, **155**
    maps between, **153**
    right, **152**
    right **cat**-, **62**, 62–65, **73**
    right deformable, **155**
    saturated, **153**
    sufficient conditions for deformability, 157
    weak equivalences between, **153**

T, 101
T-diagrams of arrow categories, **102**
terminal Kan extensions, **118**, 128, **162**, *see also* Kan extensions
terminal Kan extensions along right systems, **168**, *see also* Kan extensions along systems
terminal objects
    categorical uniqueness of, 38, 92, 113
    homotopically, 9, 13, **39**, 39–40, 83, **116**, 115–118, *see also* homotopically terminal objects
terminal projection functors, **68**, 68–72
3-arrow calculi, 6, **33**, 81, 91, **107**, 107–112
    and saturation, 11, 34, 82, 92, 110
    and 3-arrow description of homotopy categories, 5, 33, 81, 91, 109
    of model categories, 6, 34
    on homotopical diagram categories, 108
    on homotopical functor categories, 108
3-arrow description of homotopy categories, 5, 21, 32, 33, 81, 91, 109
    and 3-arrow calculi, 33
total left derived functors, 5, 7, **128**
    and left approximations, 128
total right derived functors, 5, 7, **128**
    and right approximations, 128
trivial cofibrations, **26**
    characterization of, 28
trivial fibrations, **26**
    characterization of, 28
two out of six axiom, **26**

two out of six property, 10–11, 19, **23**, 79, 90, **96**, 110, 117, 125
two out of three axiom, 19–20
two out of three property, 4, 11, **23**, 79, 90, **96**
types
   category of, **101**
types of zigzags, 81, **101**

$\mathcal{U}$-categories, **22**, 90, **94**, 94–95
   small, **22**, 90, **94**, 94–95
u-colimit functors, **57**, 85, *see also* colimit functors
u-limit functors, **57**, **148**, *see also* limit functors
$\mathcal{U}$-sets, **22**, 89, **94**
underlying categories, **96**
uniqueness
   categorical, **38**, 38–39, 92–93, **112**, 112–114, *see also* categorical uniqueness
   homotopical, 9–10, **39**, 39–40, 82–83, 92–93, **114**, 112–118, *see also* homotopical uniqueness
units
   of adjunctions, **122**, 122–123, 135–138
   of adjunctions of systems, **157**
      and dinatural transformations, 167
   of Kan extensions, **118**, **162**
universal properties, **38**, **113**
   homotopically, **40**, **115**
universes, **22**, 22–23, 78–79, 89–90, **94**, 94–95
   basic assumption, 94
   higher, **94**
   successor, **23**, 90, **94**

virtually cofibrant diagrams, **69**, 69–72
virtually fibrant diagrams, **69**, 69–72

we-categories, 4–5
weak equivalences
   between **cat**-systems, **63**
   between systems, **153**
   categories of, **96**
   in homotopical categories, 11, **23**, 90, **96**
   in model categories, 20, **25**
   in we-categories, 4
   natural, **24**, 90, **96**
   Reedy, **65**
weak invertibility property, **23**, 79, **96**
weakly equivalent objects, **23**, **96**
   canonically, **39**, 83, 92, **114**
   homotopically, 10

zigzags, **98**, 101–112
   length of, **98**
   restricted, 81, **98**, **101**
   type of, **101**

Bibliography

[AGV72] M. Artin, A. Grothendieck, and J. L. Verdier, *Theorie des topos et cohomologie étale des schémas*, Lect. Notes in Math., vol. 269, Springer-Verlag, 1972.

[BK72] A. K. Bousfield and D. M. Kan, *Homotopy limits, completions and localizations*, Lect. Notes in Math., vol. 304, Springer-Verlag, New York, 1972.

[DK80a] W. G. Dwyer and D. M. Kan, *Calculating simplicial localizations*, J. Pure Appl. Algebra **18** (1980), 17–35.

[DK80b] _____, *Function complexes in homotopical algebra*, Topology **19** (1980), 427–440.

[DK80c] _____, *Simplicial localizations of categories*, J. Pure Appl. Algebra **17** (1980), 267–284.

[DS95] W. G. Dwyer and J. Spaliński, *Homotopy theories and model categories*, Handbook of algebraic topology, North-Holland, Amsterdam, 1995, pp. 73–126.

[GJ99] P. G. Goerss and J. F. Jardine, *Simplicial homotopy theory*, Progress in Math., vol. 174, Birkhäuser Verlag, Basel, 1999.

[Hir03] P. S. Hirschhorn, *Model categories and their localizations*, Mathematical Surveys and Monographs, vol. 99, American Mathematical Society, Providence, RI, 2003.

[Hov99] M. Hovey, *Model categories*, Mathematical Surveys and Monographs, vol. 63, Amer. Math. Soc., Providence, RI, 1999.

[Mac71] S. MacLane, *Categories for the working mathematician*, Grad. Texts in Math., vol. 5, Springer-Verlag, 1971.

[Qui67] D. G. Quillen, *Homotopical algebra*, Lect. Notes in Math., vol. 43, Springer-Verlag, Berlin, 1967.

[Qui69] _____, *Rational homotopy theory*, Ann. Math. **90** (1969), 205–295.

[Ree74] C. L. Reedy, *Homotopy theory of model categories*, Available at the Hopf Topology Archive as `ftp://hopf.math.purdue.edu/pub/Reedy/reedy.dvi`, 1974, Unpublished manuscript.

[Sch72] H. Schubert, *Categories*, Springer-Verlag, 1972.

[Tho79] R. W. Thomason, *Homotopy colimits in the category of small categories*, Math. Proc. Cambridge Philos. Soc. **85** (1979), no. 1, 91–109.

Titles in This Series

113 **William G. Dwyer, Philip S. Hirschhorn, Daniel M. Kan, and Jeffrey H. Smith,** Homotopy limit functors on model categories and homotopical categories, 2004

112 **Michael Aschbacher and Stephen D. Smith,** The classification of quasithin groups II. Main theorems: The classification of simple QTKE-groups, 2004

111 **Michael Aschbacher and Stephen D. Smith,** The classification of quasithin groups I. Structure of strongly quasithin K-groups, 2004

110 **Bennett Chow and Dan Knopf,** The Ricci flow: An introduction, 2004

109 **Goro Shimura,** Arithmetic and analytic theories of quadratic forms and Clifford groups, 2004

108 **Michael Farber,** Topology of closed one-forms, 2004

107 **Jens Carsten Jantzen,** Representations of algebraic groups, 2003

106 **Hiroyuki Yoshida,** Absolute CM-periods, 2003

105 **Charalambos D. Aliprantis and Owen Burkinshaw,** Locally solid Riesz spaces with applications to economics, second edition, 2003

104 **Graham Everest, Alf van der Poorten, Igor Shparlinski, and Thomas Ward,** Recurrence sequences, 2003

103 **Octav Cornea, Gregory Lupton, John Oprea, and Daniel Tanré,** Lusternik-Schnirelmann category, 2003

102 **Linda Rass and John Radcliffe,** Spatial deterministic epidemics, 2003

101 **Eli Glasner,** Ergodic theory via joinings, 2003

100 **Peter Duren and Alexander Schuster,** Bergman spaces, 2004

99 **Philip S. Hirschhorn,** Model categories and their localizations, 2003

98 **Victor Guillemin, Viktor Ginzburg, and Yael Karshon,** Moment maps, cobordisms, and Hamiltonian group actions, 2002

97 **V. A. Vassiliev,** Applied Picard-Lefschetz theory, 2002

96 **Martin Markl, Steve Shnider, and Jim Stasheff,** Operads in algebra, topology and physics, 2002

95 **Seiichi Kamada,** Braid and knot theory in dimension four, 2002

94 **Mara D. Neusel and Larry Smith,** Invariant theory of finite groups, 2002

93 **Nikolai K. Nikolski,** Operators, functions, and systems: An easy reading. Volume 2: Model operators and systems, 2002

92 **Nikolai K. Nikolski,** Operators, functions, and systems: An easy reading. Volume 1: Hardy, Hankel, and Toeplitz, 2002

91 **Richard Montgomery,** A tour of subriemannian geometries, their geodesics and applications, 2002

90 **Christian Gérard and Izabella Łaba,** Multiparticle quantum scattering in constant magnetic fields, 2002

89 **Michel Ledoux,** The concentration of measure phenomenon, 2001

88 **Edward Frenkel and David Ben-Zvi,** Vertex algebras and algebraic curves, second edition, 2004

87 **Bruno Poizat,** Stable groups, 2001

86 **Stanley N. Burris,** Number theoretic density and logical limit laws, 2001

85 **V. A. Kozlov, V. G. Maz'ya, and J. Rossmann,** Spectral problems associated with corner singularities of solutions to elliptic equations, 2001

84 **László Fuchs and Luigi Salce,** Modules over non-Noetherian domains, 2001

For a complete list of titles in this series, visit the AMS Bookstore at www.ams.org/bookstore/.